DEDICATED TO MY MUM, LATE DEACONESS FELICIA BOLANLE ADEYEMO

Contents

INDICES, LOGARITHMS AND SURDS 3

ALGEBRAIC EQUATIONS 42

POLYNOMIALS 90

SEQUENCE AND SERIES 126

INDICES, LOGARITHMS AND SURDS

1.2. Indices

Given a number a, the product $a \times a \times a \times a \times a = a^5$, and in general,

$a \times a \times \cdots \times \cdots \times a = a^n$ is called the nth power of the number a. 'a' is called the base and n is called the Index (plural Indices).

Theorem: Let n, m be positive integers

1. $a^n \times a^m = a^{n+m}$
2. If n, m, and $a \neq 0$ then $a^n \div a^m = a^{n-m}$
3. $(a^m)^n = a^{mn}$
4. $(ab)^n = a^n \times b^n$
5. If b is a non-zero number, then $\left(\dfrac{a}{b}\right)^n = \dfrac{a^n}{b^n}$

   Proof:
   1. $a^n \times a^m$
      $= \underbrace{(a \times a \times \cdots \times a)}_{n \text{ times}} \underbrace{(a \times a \times \cdots \times a)}_{m \text{ times}}$

      $= \underbrace{a \times a \times \cdots \times a}_{(n+m) \text{ times}}$

      $= a^{n+m}$, by definition

   2. $a^n \div a^m = \dfrac{a \times a \times \cdots \times a \ (n \text{ times})}{a \times a \times \cdots \times a \ (m \text{ times})}$

      Since $n > m$, we may cancel out 'a' m times in the numerator and denominator, having
      $a \times a \times \cdots \times a \ (n - m \text{ times}) = a^{n-m}$, by definition.

   3. $(a^m)^n = \underbrace{a^m \times a^m \times \cdots \times a^m}_{n \text{ times}}$

      $= \underbrace{a^{m+m+\cdots+m}}_{n \text{ times}}$ (theorem 1)

      $= a^{mn}$

   4. $(ab)^n = \underbrace{ab \times ab \times \cdots \times ab}_{n \text{ times}}$

      $= \underbrace{(a \times a \times \cdots \times a)}_{n \text{ times}} \underbrace{(b \times b \times \cdots \times b)}_{n \text{ times}}$

      $= a^n \times b^n$, by definition

   5. $\left(\dfrac{a}{b}\right)^n = \dfrac{a}{b} \times \dfrac{a}{b} \times \cdots \times \dfrac{a}{b}$

$$= \frac{a \times a \times \cdots \times a}{b \times b \times \cdots \times b} = \frac{a^n}{b^n}$$

Corollary (to the theorem)

1. $a^0 = 1$

   Proof: In $a^n \div a^m = a^{n-m}$, let $n = m$. Then $a^n \div a^n = a^n \div a^n = a^{n-n} = a^0$
   But $a^n \div a^n = 1 \therefore a^0 = 1$

2. $a^{-n} = \frac{1}{a^n}$

   Proof: In $a^n \times a^m = a^{n+m}$
   $a^{-n} \times a^n = a^{-n+n} = a^0 = 1$

   $\therefore a^{-n} = \frac{1}{a^n}$

3. $a^{\frac{1}{n}} = \sqrt[n]{a}$

   Proof: $a^p \times a^p = a^{2p}$
   $a^p = \sqrt{a^{2p}}$
   $a^p = \sqrt[n]{a^{np}}$

   Which is an nth root of $'a'$.
   Note:

i. By substituting non-zero values for a, b, x, y and z, that $a^x = b^x$ does NOT imply $a^{x+z} = b^{x+z}$;

ii. That $a^{1/2} + b^{1/2} \neq (a+b)^{1/2}$.

Logarithms: To find the Index of an expression, given the base and value of the base to power of the index.
The Logarithm of a number y to a base b, where b is a positive number not equal to 1 is the index to which b must be raised to give y. If $\log_b y = x$, then $b^x = y$.

Note:
1. Given b and x, we obtain y by raising b to the index x.
2. Given x and y. We obtain b by taking the xth root of y (or raising y to the index $\frac{1}{x}$).
3. Given b and y, we would need the theory of logarithms to obtain x.
4. Negative numbers and zero do not have logarithms
   Logarithms to base 10 base 10 are called common logarithms.

   Note:
1. Common logarithms are used in numerical computations in science, geography, economics, commerce and industry common logarithms are simple to use because every number can be expressed in standard form. Recall that in standard form, a given number is written as a number between 1 and 10 multiplied by the appropriate index of 10 e.g., 247000000 is written in standard form as 2.47×10^8
   $\therefore \log_{10} 247000000 = 8 + \log_{10} 2.47$

2. It is clear that, to find the logarithms to base 10 of any positive real number it is sufficient to know only the logarithms of numbers between 1 and 10.
3. Moreover, if $\log_{10} x = n + \log_{10} U$ and if $1 \leq U \leq 10$, then $0 \leq \log_{10} U < 1$ and so $\log_{10} x =$ an integer n + a decimal fraction.

The integer part of $\log_{10}x$ is called the 'characteristic' and the decimal part of $\log_{10}x$ is called the 'characteristic' and the decimal part of $\log_{10}x$ is called the 'mantissa'

Properties of logarithms.

1. $\log_a(xy) = \log_a x + \log_a y$, where $a > 0, a \neq 1, x > 0, y > 0$

   Let $u = \log_a x, v = \log_a y$ and $z = \log_a(xy)$
   Then $a^z = xy = a^u \cdot a^v$
   ∴ by the law of indices
   $z = u + v$

2. $\log_b b = 1$
   $b^1 = b$ ∴ $\log_b b = 1$

3. If $\log_b y = 1$ then $y = b$
   $x = \log_b y$ f $b^x = y$

   ∴ if $\log_b y = 1$ then $y = b$

4. $\log_b 1 = 0$, for any $b > 0, b \neq 1$
   $b^0 = 1$ for any $b > 0, b \neq 1$
   $\log_b 1 = 0$, for any $b > 0, b \neq 1$

5. If $\log_b x = 0$, then $x = 1$
   If $\log_b x = 0$, then $b^0 = x$ and $x = 1$

6. $\log x_1 \cdot x_2 \cdot x_3 \cdot \ldots \cdot x_n = \log_b x_1 + \log_b x_2 + \log_b x_3 + \cdots + \log_b x_n$,

   For $b > 0, b \neq 1, x_1 > 0, x_2 > 0, x_3 > 0, \cdots, x_n > 0$.

   $\log_b x_1 \cdot x_2 \cdot x_3 \cdot \ldots \cdot x_n = z$, say
   Then $b^z = x_1 \cdot x_2 \cdot x_3 \cdot \ldots \cdot x_n$
   Let $\log_b x_1 = U_1, \log_b x_2 = U_2, \log_b x_3 = U_3, \cdots, \log_b x_n = U_n$
   Then $b^z = b^{U_1} \cdot b^{U_2} \cdot b^{U_3} \cdot \ldots \cdot b^{U_n}$
   ∴ $z = U_1 + U_2 + U_3 + \cdots + U_n$
   ∴ $\log_b x_1 \cdot x_2 \cdot x_3 \cdot \ldots \cdot x_n = \log_b x_1 + \log_b x_2 + \log_b x_3 + \cdots + \log_b x_n$
   For $b > 0, b \neq 1, x_1 > 0, x_2 > 0, x_3 > 0, \cdots, x_n > 0$.

   Note: Even on the condition that $x_1 > 0, x_2 > 0, x_3 > 0, \cdots, x_n > 0$. The product of two negative numbers is meaningful, (since it is a positive number), but then $\log_b(x_1 x_2) = \log_b[(-x_1)(-x_2)]$.

7. $\log_b \frac{1}{x} = -\log_b x$.

   Let $v = \log_b \frac{1}{x}$. Then $b^v = \frac{1}{x}$ or $b^{-v} = x$

   ∴ $\log_b \frac{1}{x} = -\log_b x$.

8. $\log_b \frac{x}{y} = \log_b x - \log_b y$,
   $b > 0, b \neq 1, x > 0, y > 0$

$$\log_b \frac{x}{y} = \log_b \left(x \times \frac{1}{y}\right)$$

$$\log_b \left(x \cdot \frac{1}{y}\right) = \log_b x + \log_b \frac{1}{y} \text{ (property 1)}$$

Since $\log_b \frac{1}{y} = -\log_b y$ (property 7)

Then $\log_b x + \log_b \frac{1}{y} = \log_b x - \log_b y$

$\therefore \log_b \frac{x}{y} = \log_b x - \log_b y$.

9. $\log_b x^m = m\log_b x$, $b > 0, b \neq 1, x > 0$, m may be positive or negative.

    Let $\log_b x^m = z$, then $b^z = x^m$

    $b^{z \times \frac{1}{m}} = x$; $b^{\frac{z}{m}} = x$

    $\frac{z}{m} = \log_b x$

    $\frac{z}{m} \times m = m\log_b x$

    $\therefore z = m\log_b x$

    i.e. $\log_b x^m = m\log_b x$

10. $\log_{b^n} y = \frac{1}{n}\log_b y$

    Let $v = \log_{b^n} y$

    Then $(b^n)^v = y$; $b^{nv} = y$

    $b^v = y^{\frac{1}{n}}$

    $\therefore v = \log_b y^{\frac{1}{n}} = \frac{1}{n}\log_b y$

11. $\log_{b^n} y^n = \log_b y$

    Since $\log_{b^n} y = \frac{1}{n}\log_b y$ and $\log_b y^n = n\log_b y$ then $\log_{b^n} y^n = n \times \frac{1}{n}\log_b y$

    $\log_{b^n} y^n = \log_b y$

12. $\log_b y = \frac{\log_a y}{\log_a b}$

    Let $v = \log_b y$. Then $b^v = y$.

    Taking logarithms to base a, $\log_a b^v = \log_a y$
    $v\log_a b = \log_a y$

    $v = \log_b y = \frac{\log_a y}{\log_a b}$

13. $\log_b a = \frac{1}{\log_a b}$, $a > 0, a \neq 1, b > 0, b \neq 1$

Let $\log_b a = v$

Then $b^v = a$

Taking logarithms to base a

$\log_a b^v = \log_a a$

$v \log_a b = 1$

$v = \dfrac{1}{\log_a b}$

14. $\log_b \sqrt[n]{x} = \dfrac{1}{n} \log_b x$

Since $\sqrt[n]{x} = x^{\frac{1}{n}}$ then $\log_b \sqrt[n]{x} = \log_b x^{\frac{1}{n}}$

$\log_b \sqrt[n]{x} = \dfrac{1}{n} \log_b x$ (property 9)

15. If b and y are positive and $b \neq 1, y \neq 1$, then $\log_b y \times \log_y b = 1$.

Since $\log_b y = \dfrac{1}{\log_y b}$ then $\dfrac{1}{\log_y b} \times \dfrac{\log_y b}{1} = 1$

Valuable Examples

1. $9 \log_a 5 = \log_5 a$

$\dfrac{9}{\log_5 a} = \log_5 a$

$\therefore 9 = (\log_5 a)^2$

$9^{\frac{1}{2}} = \log_5 a$

$\log_5 a = \pm \sqrt{9}$

$\log_5 a = -3$ or $\log_5 a = 3$

$a = 5^{-3}$ or $a = 5^3$

$a = \dfrac{1}{125}$ or $a = 125$

2. If $2 = 10^{0.3010}$ and $3 = 10^{0.4771}$

Find $\log_{10} 6$

$\log_{10} 6 = \log_{10}(2 \times 3) = \log_{10} 2 + \log_{10} 3$
$= 0.3010 + 0.4771$
$= 0.7781$

Using the logarithm table

3. $\log_{10} 0.2$

$\log_{10} 0.2 = \log_{10}(10^{-1} \times 2)$
$= -1 + 0.3010$
$= \bar{1}.3010$ (pronounced "bar 1.3010)

4. Use log tables to find $\log_{10} 29.74$

First put the number in standard form, 2.974×10

The characteristic (index of 10) is 1. That's because the actual value, 29.74 is in tens (10^1). And that the ten in the standard form is to power 1 i.e. 2.974×10^{-1}.

To check this number under logarithms, check the first 2 digits in the first tab of the four-figure table, then you check the 'number' that is horizon tally beside the 2 digits number and at the same time under the third-digit number i.e. the 'number' must be under the third-digit column and on the roll of the 2-digit number must be added from the 'number' gotten in the third-digit column and on the 2-digit roll. The final number gotten after the addition is the logarithm of the number in question. All these are checked in the table of logarithm of number.

$\log_{10} 29.74$

Convert to standard from $\log_{10} 2.974 \times 10$
$\log_{10} 10 + \log_{10} 2.974 = 1 + \log_{10} 2.974$

(check the number in column '7' on the roll where you have '29')

The number is 4728

(check the number in the column 4 on the difference tab the same roll)

The number is 6
(The difference is added to the first number derived)

$4728 + 6 = 4734$

$\therefore \log_{10} 2.974 = 0.4734$

$\log_{10} 29.74 = 1 + 0.4734$
$= 1.4734$

5. Use antilog tables to find the number 'a'
   Such that
   $\log_{10} a = \bar{3}.4215$.

   Ignore the characteristic (3).
   Look up $\cdot 4215$ in antilog tables, giving 2639. Then $a = 2.639 \times 10^{-3}$
   $\therefore a = 0.002639$

6. Using log tables, calculate
   a. 89.31×0.6218
   b. $0.07304 \div 0.8931$

   Logs are added in other to multiply them and subtracted to divide

   a. | Number | Log |
      |---|---|
      | 89.31 | 1.9509 |
      | 0.6218 | $\bar{1}.7937$ |
      | | 1.7446 |

   Using antilog tables, the solution is $5.554 \times 10 = 55.54$

   b. | Number | Log |
      |---|---|
      | 0.07304 | $\bar{2}.8636$ |
      | 0.8931 | $\bar{1}.9509$ |
      | | $\bar{2}.9127$ |

   Using antilog tables, the solution is $8.179 \times 10^{-2} = 0.08179$

7. Find $\sqrt{29.5}$
   Let $y = \sqrt{29.5}$
   $\log_{10} y = \log_{10} (29.5)^{\frac{1}{2}} = \frac{1}{2} \log_{10} 29.5$

Using log tables
$$\log_{10} y = \frac{1}{2}(1.4698) = 0.7349$$

Using log tables.

$$\log_{10} y = \frac{1}{2}(1.4698) = 0.7349$$

Using antilog tables
$y = 5.431$
$\therefore \sqrt{29.5} = 5.4$

$\therefore$ All mathematical problems involving decimal can be solved using the logarithm and anti-logarithm tables

8. Carbon-14 is a radioactive isotope of carbon which has a half-life of about 5600 years (that is, of a given quantity of carbon 15 about half would decay in 5600 years). Let R be the nearly constant ratio of carbon-15 to carbon-12 (this is the non-radioactive isotope which is the form of most of atmosphere carbon) found in the atmosphere. Let r be the ratio of carbon-14 to carbon-12 found in an observed specimen. It has been shown for carbon-14 dating of objects, that
$R = re^{(t \ln 2)/5600}$

Where t is the age of the object in years. Suppose a specimen has been found in which $r = 0.1R$, find the age of the specimen
PLS CHECK

$$\frac{R}{r} = e^{(t \ln 2)/5600}$$

$$\frac{R}{0.1R} = e^{(t \ln 2)/5600}$$

$10 = e^{(t \ln 2)/5600}$
$\ln 10 = e^{(t \ln 2)/5600}$

$\ln 10 = \ln e^{(t \ln 2)/5600}$
$\ln 10 = \left[\frac{t \ln 2}{5600}\right] \ln e$

Since $\ln e = 1$
$\therefore \ln 10 = \frac{t \ln 2}{5600}$

Multiply both sides by $\frac{5600}{\ln 2}$

$$\frac{5600}{\ln 2} \cdot \frac{\ln 10}{1} = t$$

From the tables of natural logarithms,
$\ln 2 = 0.6931$, $\ln 10 = 2.3026$
$$\therefore t = \frac{5600(2.3026)}{0.6931}$$

$= 18,600$ yrs

9. Evaluate $(0.32)^{0.41}$

No	Standard Form	Log
0.32	3.2×10^{-1}	$\bar{1}.5051$

Now $\bar{1}.5051 = -1 + 0.5051$
$= -0.4949$

$$-0.4949 \times 0.41$$
$$= -0.202909$$
$$= -1 + 0.797091$$
$$= -1 + 0.7971$$

$\therefore \log(0.32)^{0.41} = (0.41)\log(0.32) = \bar{1}.7971$

Check the antilog tables for 0.7971 and place the decimal using $\bar{1}$
Ans: 6.27×10^{-1}, to 2 decimal places $= 0.63$.

*

Natural (or Nepirian) Logarithm
In calculus, a man called John Napier discovered that the most convenient base to use is a number denoted by $e \approx 2.718$.
Thus logarithms to base e are called natural or Napierian logarithms and are denoted by $\log_e$ or $\ln$.

SURDS

A number which can be expressed as the quotient $\dfrac{m}{n}$ of two integers, $(n \neq 0)$, is called a rational number. Any real number which is not rational is called irrational. Irrational numbers which are in the form of roots are called surds.

$\sqrt{2}, \sqrt{3}, \sqrt{5}, \pi$ and $\sqrt[3]{2}$ are all irrational numbers, while $\sqrt{16}, \sqrt[3]{8}$ and $\sqrt[5]{32}$ are not surds since they can be simplified to obtain rational numbers 4, 2 and 2, respectively. A general surd is an irrational number of the form $a\sqrt[n]{b}$, where a is a rational number and $\sqrt[n]{b}$ is an irrational number. $\sqrt[n]{b} = b^{\frac{1}{n}}$ and $\sqrt[n]{}$ is called a RADICAL.

In the determination of the value of $\dfrac{1}{\sqrt{2}}$ as a decimal, for instance, it is obviously easier to perform the division $\dfrac{1.414}{2}$ and $\dfrac{1}{1.414}$.

The simplification of $\dfrac{1}{\sqrt{2}}$ to $\dfrac{1}{\sqrt{2}} \times \dfrac{\sqrt{2}}{\sqrt{2}} = \dfrac{\sqrt{2}}{2}$ is known as RATIONALIZATION of the denominator, since the denominator is now rational.

Note: $\sqrt{b}$ is taken to mean the positive square root of b, the negative square of b, the negative square of b is $-\sqrt{b}$.

Rules for manipulating surds

1. $a\sqrt{b} + c\sqrt{b} = (a+c)\sqrt{b}$ (addition law of surds with the same radical).

2. $a\sqrt{d} - c\sqrt{d} = (a-c)\sqrt{d}$
   (Subtraction law)

3. $\sqrt{ab} = \sqrt{a} \cdot \sqrt{b}$

4. $(a\sqrt{b}) \cdot (c\sqrt{d}) = ac\sqrt{bd}$

5. $\sqrt{\dfrac{a}{b}} = \dfrac{\sqrt{a}}{\sqrt{b}}$

6. $(a\sqrt{b}) \div (c\sqrt{d}) = \dfrac{a}{c}\sqrt{\dfrac{b}{d}}$

7. $(\sqrt{a})^2 = a = \sqrt{a^2}$

8. $(\sqrt{a})^n = \sqrt{a^n}$

9. $\sqrt{a^{-m}} = \dfrac{1}{\sqrt{a^m}}$ and $\dfrac{1}{\sqrt{a^{-m}}} = \sqrt{a^m}$

Note:

1. (3) – (9) above follow from the laws of indices

2. $a\sqrt{b} + c\sqrt{d},\ b \ne d$

   $a\sqrt{b} - c\sqrt{d},\ b \ne d$

   Cannot be simplified further.

   Useful hints on rationalization
   1. $\sqrt{a} \cdot \sqrt{a} = a$
   2. $(\sqrt{a} + \sqrt{b})(\sqrt{a} - \sqrt{b}) = a - b$ (difference of two square)
   3. $(x\sqrt{a} + y\sqrt{b})(x\sqrt{a} - y\sqrt{b}) = x^2 a - y^2 b$
   4. $(x + y\sqrt{b})(x - y\sqrt{b}) = x^2 - y^2 b$

   All yield rational numbers. Therefore, the surds $\sqrt{a}$ and $-\sqrt{a}$ are called conjugate surds. In the same way, $\sqrt{a} + \sqrt{b}$ and $\sqrt{a} - \sqrt{b}$, are conjugate surds. Similarly, $x + y\sqrt{b}$, and $x - y\sqrt{b}$, are conjugate pairs of surds.

   These are very useful in rationalizing

   Note: Squaring an equation alters the equation. If two answers are derived from the equation, it is possible that, after squaring the equation, only one of the answers is correct. Finding the square root of a $a + c\sqrt{d},\ a - c\sqrt{d}$.

   Theorem:
   Let $\sqrt{a + c\sqrt{d}}$ be $\sqrt{x} + \sqrt{y}$.
   Then $\sqrt{a + c\sqrt{d}} = \sqrt{x} + \sqrt{y}$.

   Then, squaring both sides, we obtain $a + c\sqrt{d} = x + y + 2\sqrt{xy}$ \hfill (1)

   But $(\sqrt{x} - \sqrt{y})^2 = x + y - 2\sqrt{xy}$

   From (1) we have
   $a = x + y,\ d = xy$ and $c = 2$
   $\therefore (\sqrt{x} - \sqrt{y})^2 = a - c\sqrt{d}$

Exercise 1.1
Show that

1. $a^{\frac{1}{2}} \times b^{\frac{1}{2}} = (ab)^{\frac{1}{2}}$
   According to theorem 4
   Since $(ab)^{\frac{1}{2}} = a^{\frac{1}{2}} \times b^{\frac{1}{2}}$
   Then $a^{\frac{1}{2}} \times b^{\frac{1}{2}} = (ab)^{\frac{1}{2}}$

2. Show that
   $(a^x)^{m/n} = \sqrt[n]{a^{xm}}$ for any indices $x, \dfrac{m}{n}$.

   $(a^x)^{m/n} = a^{\frac{x}{1} \cdot \frac{m}{n}} = a^{xm/n}$ (Theorem 3)

   Since $a^{\frac{1}{2}} = \sqrt[2]{a^1}$ then $a^{xm/n} = \sqrt[n]{a^{xm}}$

Exercise 1.2

1. Simplify $\dfrac{4a^2b}{\sqrt[3]{2ac}}$

$$\dfrac{4a^2b}{(2ac)^{\frac{1}{3}}}$$

Multiply numerator and denominator by $\sqrt[3]{4a^2c^2}$

$$\dfrac{4a^2b}{\sqrt[3]{2ac}} \cdot \dfrac{\sqrt[3]{4a^2c^2}}{\sqrt[3]{4a^2c^2}}$$

$$\dfrac{4a^2b \cdot \sqrt[3]{4a^2c^2}}{\sqrt[3]{2 \times 4 \times a \times a^2 \times c \times c^2}}$$

$$\dfrac{4a^2b \cdot \sqrt[3]{4a^2c^2}}{\sqrt[3]{8a^3c^3}}$$

$$\dfrac{4a^2b \cdot \sqrt[3]{4a^2c^2}}{2ac}$$

$$\dfrac{2ab}{c} \cdot \sqrt[3]{4a^2c^2}$$

2. Rationalize $\dfrac{2}{\sqrt{5} - \sqrt{3}}$

$$\dfrac{2}{\sqrt{5} - \sqrt{3}} \cdot \dfrac{\sqrt{5} + \sqrt{3}}{\sqrt{5} + \sqrt{3}}$$

$$\dfrac{2(\sqrt{5} + \sqrt{3})}{5 + \sqrt{15} - \sqrt{15} - 3}$$

$$= \sqrt{5} + \sqrt{3}$$

3. Express $\dfrac{7 - 4\sqrt{5}}{3\sqrt{5} + 5\sqrt{3}}$ in the form of $m\sqrt{5} + n\sqrt{3}$, where m and n are rational numbers.

$$\dfrac{7 - 4\sqrt{5}}{3\sqrt{5} + 5\sqrt{3}} \cdot \dfrac{3\sqrt{5} - 5\sqrt{3}}{3\sqrt{5} - 5\sqrt{3}}$$

$$\dfrac{21\sqrt{5} - 35\sqrt{3} - 12.5 + 20\sqrt{15}}{9.5 - 15\sqrt{16} + 15\sqrt{15} - 25.3}$$

$$= \dfrac{21\sqrt{5} - 35\sqrt{3} + 20\sqrt{15} - 60}{45 - 70}$$

$$= \dfrac{21\sqrt{5} - 35\sqrt{3} + 20\sqrt{15} - 60}{-30}$$

$$= -\dfrac{21\sqrt{5}}{30} + \dfrac{35\sqrt{3}}{30} - \dfrac{20\sqrt{5}}{30} + \dfrac{60}{30}$$

$$= -\dfrac{7\sqrt{5}}{10} + \dfrac{7\sqrt{3}}{6} - 2\left(\dfrac{\sqrt{15}}{3} - \dfrac{1}{1}\right)$$

$$= 7\left(\dfrac{\sqrt{3}}{6} - \dfrac{\sqrt{5}}{10}\right) - 2\left(\dfrac{\sqrt{15} - 3}{3}\right)$$

$$= \frac{7}{6}\sqrt{3} - \frac{7}{10}\sqrt{5} - 2\left(\frac{\sqrt{15}-3}{3}\right)$$

4. Express $\dfrac{3-\sqrt{5}}{3+\sqrt{5}}$ in the form $a + b\sqrt{c}$, where a and b are rational numbers.

$$\frac{3-\sqrt{5}}{3+\sqrt{5}} \cdot \frac{3-\sqrt{5}}{3-\sqrt{5}}$$

$$= \frac{9 - 3\sqrt{5} - 3\sqrt{5} + 5}{9 - 5}$$

$$= \frac{9 - 6\sqrt{5} + 5}{4}$$

$$= \frac{14 - 6\sqrt{5}}{4}$$

$$= \frac{14}{4} - \frac{6}{4}\sqrt{5}$$

$$= \frac{7}{2} - \frac{3}{2}\sqrt{5}$$

5. Expand $(2\sqrt{3} - 1)^2$

$(2\sqrt{3} - 1)(2\sqrt{3} - 1)$

$4 \cdot 3 - 2\sqrt{3} - 2\sqrt{3} + 1$

$12 + 1 - 4\sqrt{3}$

$= 13 - 4\sqrt{3}$

6. Verify that $x = 1 + \sqrt{3}$ and $x = 1 - \sqrt{3}$ are solutions of the equation $x^3 - x^2 - 4x - 2 = 0$

Value worth testing are the factors of

$$-\frac{2}{1} = -2$$

They are: $-1, 1, 2, -2$

When $f(-1) = -1 - 1 + 4 - 2 = 0$

$f(1) = 1 - 1 - 4 - 2 = -6$

$f(2) = 8 - 4 - 8 - 2 = -6$

$f(-2) = -8 - 4 + 8 - 2 = -6$

$\therefore x = -1$ i.e. $(x + 1)$ is a factor.
$\therefore$

```
              x² - 2x - 2
        ┌─────────────────────
x + 1   │  x³ - x² - 4x - 2
        │  - (x³ + x²)
        │  ─────────────
        │       - 2x² - 4x - 2
        │       - (- 2x² - 2x)
        │       ─────────────
        │              - 2x - 2
        │              - (- 2x - 2)
        │              ─────────────
```

Find the roots of $x^2 - 2x - 2 = 0$

Solving by formula method

$$\frac{-b \pm \sqrt{b^2 - 4ac}}{2a} = x$$

Where $a = 1, b = -2$ and $c = -2$

$$\frac{-(-2) \pm \sqrt{4 - 4(-2)}}{2} = x$$

$$\frac{2 \pm \sqrt{4 + 8}}{2} = x$$

$$x = \frac{2 + \sqrt{12}}{2}$$

or $x = \frac{2 - \sqrt{12}}{2}$

$x = \frac{2(1 + \sqrt{3})}{2}$ or $x = \frac{2(1 - \sqrt{3})}{2}$

$x = 1 + \sqrt{3}$ or $x = 1 - \sqrt{3}$

7. Find the two smallest integers so that the value of the surd is between them
   a. $\sqrt{43}$
   b. $\sqrt{91}$
   c. $\sqrt{99}$
   d. $\sqrt{29}$
   e. $\sqrt{135}$
   f. $\sqrt{37}$

   a. $6 < \sqrt{43} < 7$
   b. $9 < \sqrt{91} < 10$
   c. $9 < \sqrt{99} < 10$
   d. $5 < \sqrt{29} < 6$
   e. $11 < \sqrt{135} < 12$
   f. $6 < \sqrt{37} < 7$

Exercise 1.3

1. Evaluate $\sqrt{24 + 6\sqrt{15}}$

   Let $\sqrt{24 + 6\sqrt{15}} = \sqrt{x} + \sqrt{y}$ (1)
   Squaring both sides we have.

   $24 + 6\sqrt{15} = x + y + 2\sqrt{xy}$ then $x + y = 24$ (3)

   Then $\sqrt{24 - 6\sqrt{15}} = \sqrt{x} + \sqrt{y}$ (2)

   Multiply (1) and (2)
   $\sqrt{24^2 - 36 \cdot 15} = x - y$
   $x - y = \sqrt{576 - 540}$
   $x - y = \sqrt{36}$

   $x - y = 6$ (4)
   Add (3) & (4)

   $2x = 30$
   $x = 15$

Then $y = 9$

$\therefore \sqrt{24 + 6\sqrt{15}} = \sqrt{15} + \sqrt{9}$
$= 3 + \sqrt{15}$

2. Show that
$\sqrt{5 + \sqrt{15} + (\sqrt{24 + 6\sqrt{15}})} = \sqrt{3} + \sqrt{5}$

Since $\sqrt{24 + 6\sqrt{5}} = \sqrt{15} + \sqrt{9}$

Then $\sqrt{3 + \sqrt{15} + (\sqrt{15} + \sqrt{9})}$

$\sqrt{5 + 2\sqrt{15} + 3}$

$\sqrt{8 + 2\sqrt{15}}$

Let $\sqrt{8 + 2\sqrt{15}} = \sqrt{x} + \sqrt{y}$

Squaring both sides, we have
$8 + 2\sqrt{15} = x + y + 2\sqrt{xy}$

$x + y = 8; \ \& \ xy = 15$

By Inspection, $x = 3 \text{ and } y = 5$

$\therefore \sqrt{5 + \sqrt{15} + (\sqrt{24 + 6\sqrt{15}})} = \sqrt{3} + \sqrt{5}$

3. Rationalize $\dfrac{1}{\sqrt{4 - \sqrt{15}}}$

Find the roots of $4 - \sqrt{15}$ first

Let $\sqrt{4 - \sqrt{15}} = \sqrt{x} - \sqrt{y}$ \quad (1)

Then $\sqrt{4 + \sqrt{15}} = \sqrt{x} + \sqrt{y}$ \quad (2)

$\sqrt{4 - \sqrt{15}} = x + y - 2\sqrt{xy}$ then $x + y = 4$ (3)

Multiply (1) & (2)
$\sqrt{16 - 15} = x - y; \ x - y = \sqrt{1};$

$x - y = 1$ (4)

Add (3) & (4)
$2x = 5$
$x = \dfrac{5}{2}$

Then $y = \dfrac{4}{1} - \dfrac{5}{2} = \dfrac{8 - 5}{2}$

$y = \dfrac{3}{2}$

Back to Question:
$\dfrac{1}{\sqrt{\dfrac{5}{2}} - \sqrt{\dfrac{3}{2}}} = \dfrac{1}{\dfrac{\sqrt{5} - \sqrt{3}}{\sqrt{2}}}$

$1 \times \dfrac{\sqrt{2}}{\sqrt{5} - \sqrt{3}}$

$$\frac{\sqrt{2}}{\sqrt{5}-\sqrt{3}} \cdot \frac{\sqrt{5}+\sqrt{3}}{\sqrt{5}+\sqrt{3}}$$

$$\frac{\sqrt{10}+\sqrt{6}}{25-9} = \frac{\sqrt{10}+\sqrt{6}}{16}$$

4. Simplify $\dfrac{4}{1+\sqrt{2}-\sqrt{3}}$

$$\frac{4}{1+\sqrt{2}-\sqrt{3}} \times \frac{1+\sqrt{2}+\sqrt{3}}{1+\sqrt{2}+\sqrt{3}}$$

$$\frac{4-4\sqrt{2}+4\sqrt{3}}{1-\sqrt{2}+\sqrt{3}+\sqrt{2}-\sqrt{2}+\sqrt{6}-\sqrt{3}+\sqrt{6}-3}$$

$$\frac{4(1-\sqrt{2}+\sqrt{3})}{2\sqrt{6}-4}$$

$$\frac{4(1-\sqrt{2}+\sqrt{3})}{2(\sqrt{6}-2)}$$

$$\frac{2-2\sqrt{2}+2\sqrt{3}}{\sqrt{6}-2} \cdot \frac{\sqrt{6}+2}{\sqrt{6}+2}$$

$$\frac{2\sqrt{6}+4-2\sqrt{6}-4\sqrt{2}+6\sqrt{2}+4\sqrt{3}}{2}$$

$$\frac{4+2\sqrt{6}+2\sqrt{2}}{2}$$

$$2+\sqrt{6}+\sqrt{2}$$

$$= 2+\sqrt{2}+\sqrt{6}$$

5. If $x = \dfrac{1}{2}(1-\sqrt{5})$, express $4x^3 - 3x$ in its simplest surd form.

$$4\left(\frac{1-\sqrt{5}}{2}\right)^3 - 3\left(\frac{1-\sqrt{5}}{2}\right)$$

$$4\left(\frac{(1-\sqrt{5}-\sqrt{5}+5)(1-\sqrt{5})}{2^3}\right) - 3\left(\frac{(1-\sqrt{5})}{2}\right)$$

$$4\left(\frac{1-2\sqrt{5}+5-\sqrt{5}+2.5-5\sqrt{5}}{2}\right) - 3\left(\frac{1-\sqrt{5}}{2}\right)$$

$$\frac{(16-8\sqrt{5})}{2} - \left(\frac{3-3\sqrt{5}}{2}\right)$$

$$\frac{16-8\sqrt{5}-3+3\sqrt{5}}{2}$$

$$\frac{13-5\sqrt{5}}{2}$$

$$= \frac{1}{2}(13-5\sqrt{5})$$

Exercise 1.4

1. Simplify the following:
   a. $\log_2 12$
   b. $\log_3 1/9$
   c. $\log_4 8$

   a. $\log_2 12 = \log_2 (3 \times 4)$
   $= \log_2 3 + \log_2 2^2$
   $= \log_2 3 + \log_2 2^2$
   $= \log_2 3 + 2\log_2 2$
   $= \log_2 3 + 2$

   b. $\log_3 1/9 = \log_3 9^{-1}$
   $= -1\log_3 3^2 = -1 \times 2\log_3 3$
   $= -2$

   c. $\log_4 8 = \log_{2^2} 2^3$
   $= \frac{1}{2} \times 3\log_2 2$
   $= 3/2$

2. Evaluate

   $\left(2\log 7 + 3\log \frac{14}{15} + 5\log \frac{3}{7} + \log \frac{11}{8}\right) - \log \frac{99}{125}$

   $\left(\log 7^2 + \log \left(\frac{14}{15}\right)^3 + \log \left(\frac{3}{7}\right)^5 + \log \frac{11}{8}\right) - \log \frac{99}{125}$

   $\log \left(7^2 \times \left(\frac{14}{15}\right)^3 \times \frac{3^5}{7^5} \times \frac{11}{8}\right) - \log \frac{99}{125}$

   $\log \left(\frac{14^3}{15^3} \times \frac{3^5}{7^3} \times \frac{11}{8}\right) - \log \frac{99}{125}$

   $\log \left(\frac{1}{5^3} \times \frac{3^2}{1} \times \frac{11}{1}\right) - \log \frac{99}{125}$

   $\log \frac{11 \times 9}{5^3} - \log \frac{99}{125}$

   $\log \left(\frac{99}{125} \times \frac{125}{99}\right)$

   $\log 1 = \log_{10} 1$

   $= 0$

3. Evaluate

   $81^{\frac{1}{\log_2 3}} \div 36^{\frac{1}{\log_2 6}}$

$$81^{\frac{\log_2 2}{\log_2 3}} \div 36^{\frac{\log_2 2}{\log_2 6}}$$

$$81^{\log_3 2} - 36^{\log_6 2} \quad \text{(Theorem 13)}$$

$$3^{4\log_3 2} \div 6^{2\log_6 2}$$

Let $3^{4\log_3 2}$ be x and $6^{2\log_6 2}$ by y

$x \div y$

i.e.
$3^{4\log_3 2} = x \qquad 6^{2\log_6 2} = y$

Add $\log_3$ to both sides   Add $\log_6$ to both sides
$\log_3 3^{4\log_3 2} = \log_3 x \qquad \log_6 6^{2\log_6 2} = \log_6 y$
$4\log_3 2 \times \log_3 3 = \log_3 x \qquad 2\log_6 2 \times \log_6 6 = \log_6 y$

$4\log_3 2 = \log_3 x \qquad 2\log_6 2 = \log_6 y$

$\log_3 2^4 = \log_3 x \qquad \log_6 2^2 = \log_6 y$

$\log_3 16 = \log_3 x \qquad \log_6 4 = \log_6 y$

$\therefore 16 = x \qquad \therefore 4 = y$

i.e.
$x \div y = 16 \div 4$
$= 4$

Alternative Method:

$$81^{\frac{1}{\log_2 3}} \div 36^{\frac{1}{\log_2 6}}$$

$$81^{\log_3 2} \div 36^{\log_6 2} \quad \text{(Theorem 13)}$$

$$3^{4\log_3 2} \div 6^{2\log_6 2}$$

$$3^{4 \times \log_3 2} \div 6^{2 \times \log_6 2}$$

$$3^{4\left[\frac{\log_{10} 2}{\log_{10} 3}\right]} \div 6^{2\left[\frac{\log_{10} 2}{\log_{10} 6}\right]} \quad \text{(Theorem 12)}$$

$3^{4(0.63089499)} \div 6^{2(0.38679003)}$

$3^{2.52358} \div 6^{0.77358006}$

$16 \div 4$

= 4

4. Evaluate
   a. $\log_4 \dfrac{1}{8}$
   b. $\log_{\frac{1}{3}} 27$
   c. $\log_2 11 \times \log_6 2$

   a. $\log_4 \dfrac{1}{8} = \log_{2^2} 8^{-1}$

   $= \log_{2^2} 2^{-3}$

   $= \dfrac{1}{2} \times -3 \log_2 2 = -\dfrac{3}{2}$

   b. $\log_{\frac{1}{3}} 27 = \log_{3^{-1}} 3^3$

   $= -\dfrac{1}{1} \times 3 \times \log_3 3$

   $= -3$

   c. $\log_2 11 \times 1/\log_2 6$

   $= \dfrac{\log_2 11}{\log_2 6} = \log_6 11$

5. Simplify
   $\dfrac{1}{\log_a ab} + \dfrac{1}{\log_b ab}$

   $= \log_{ab} a + \log_{ab} b$

   $= \log_{ab} a \times b$

   $= \log_{ab} ab = 1$

6. If $a = \log_b c,\ b = \log_c a,\ c = \log_a b$
   Prove that
   $abc = 1$

   $abc = a \times b \times c$

   $= \log_b c \times \log_c a \times \log_a b$

   $\log_b c \times \dfrac{1}{\log_a c} \times \dfrac{\log_a b}{1}$

   $\log_b c \times \dfrac{\log_a b}{\log_a c}$

   $= \log_b c \times \log_c b = \log_b c \times \dfrac{1}{\log_b c}$

   $= 1$

7. If a, b, c are positive numbers show that
$$(1 + \log_a b)\log_{ab} c = \log_a c$$

$$(\log_a a + \log_a b)\log_{ab} c$$

$$= \log_a ab \times \log_{ab} c$$

$$= \frac{1}{\log_{ab} a} \times \log_{ab} c$$

$$= \frac{\log_{ab} c}{\log_{ab} a} = \log_a c$$

8. By changing to base 10, show that $\log_3 20.37 \cong 2.7435$

$$\log_3 20.37 = \frac{\log_{10} 20.37}{\log_{10} 3} \quad \text{(Theorem 12)}$$

$$= \frac{1.3090}{0.4771} = 2.7436596$$

$$\therefore \log_3 20.37 \cong 2.7437$$

9. Find a if $\log_{10} a^2 = \bar{1}.5784$

$$\log_{10} a^2 = \bar{1}.5784$$

$$2\log_{10} a = \bar{1}.5784$$

$$\log_{10} a = \frac{\bar{1}.5784}{2}$$

$$\log_{10} a = \frac{\bar{2} + 1.5784}{2}$$

$$\log_{10} a = \bar{1} + 0.7892$$

$$\log_{10} a = \bar{1}.7892$$

$$a = 10^{\bar{1}.7892}$$

$$a = 0.6155$$

10. The number of bacteria in a culture at time t is given by $n = 2.000e^{5t}$. Wat is the value of t when the colony is double its initial size?

$$n = 2000e^{5t}$$

If the colony has doubled then the time has also doubled.
i.e.
1. $n = 2000e^{5t}$ (the initial size)
2. $2n = 2000e^{5 \times 2t}$ (the new size)

Which means:

$$\frac{2000e^{5 \times 2t}}{2} = \frac{2(2000e^{5t})}{2}$$

$$\frac{1000e^{10t}}{1000} = \frac{2000e^{5t}}{1000}$$

$$e^{10t} = 2e^{5t}$$

Add $\log_e$ to both sides

$$\log_e e^{10t} = \log_e 2e^{5t}$$

$$10t\log_e e = \log_e 2 + \log_e e^{5t}$$

$$10t = \ln 2 + 5t\log_e e$$

$$10t = 0.6931 + 5t$$

$$10t - 5t = 0.6931$$

$$\frac{5t}{5} = \frac{0.6931}{5}$$

$$t = 0.13862$$

$$t \cong 0.139$$

11. y m³ of a water measured as flowing per second over a dam when the difference of water levels was x m. The following result were obtained:

x	1.2	1.4	1.6	1.8	2.0
y	6.3	9.2	12.8	17.5	22.4

Plot a graph of $\log_{10} y$ against $\log_{10} x$. Show that $y = ax^n$ and by reading off the slope and intercept, find approximate values of n and $\log_{10} a$ respectively, and hence find approximate values of the constants a and n.

Tables:

x	1.2	1.4	1.6	1.8	2.0
$\log_{10} x$	0.0792	0.1461	0.2041	0.2553	0.3010
y	6.3	9.2	12.8	17.5	22.4
$\log_{10} y$	0.7993	0.9638	1.1072	1.2380	1.3502

Graph: PLS CHECK

Whatever the equation, it is clear that when the level of the water, x, is 0, the volume of water flowing per second, y, will also be 0. This shows that the equation that shows the relationship between x and y is not a quadratic or cubic or more equation.

i.e.
$$y \neq ax^n + bx^{n-1} + cx^{n-2}$$

Since the equation has only one root which is 0 i.e. when $x = 0$; $y = 0$.

So also the equation is not a linear one because the ratio of y to x is not constant and the curve is not a linear curve (i.e. a straight line).

$$y \neq ax + b$$

Since the equation has only one root and therefore it is neither a linear nor quadratic (nor cubic or more) equation, then $y = ax^n$ when n is not a whole number

Because
1. The equation has only one root
2. When $x = 0, y = 0$ too.

Where n is not a whole number

$\therefore y = ax^n$

$\log_{10} y = \log_{10} a \cdot x^n$

$\log_{10} y = \log_{10} a + n\log_{10} x$

$0.7993 = \log_{10} a + n \times 0.0792$ \quad (1)

$0.9638 = \log_{10} a + n \times 0.1461$ \quad (2)

$0.0792n + \log_{10} a = 0.7993$ \quad (3)

$0.1461n + \log_{10} a = 0.9638$ \quad (4)

(4) ÷ (3)

$\dfrac{0.0669n}{0.0669} = \dfrac{0.1645}{0.0669}$

$n = 2.4588938$

$n \cong 2.46$

In equ. (3)

$0.0792n + \log_{10} a = 0.7993$

$0.0792 \times 2.4588938 + \log_{10} a = 0.7993$

$0.1947443 + \log_{10} a = 0.7993$

$\log_{10} a = 0.7993 - 0.1947443$

$\log_{10} a = 0.6046$

$a = 10^{0.6046}$

$a = 4.024$

$a \cong 4.0$

12. Given an integer n, show that $(-a)^n = a^n$ if n is even $= -(a^n)$ if n is odd

    Assume $n = 2, 4, 6, 8$

    If $n = 2$: $(-a)^2 = -a \times -a = a^2$
    $(-a)^4 = (-a \times -a)(-a \times -a) = a^4$
    $(-a)^6 = (-a \times -a)(-a \times -a)(-a \times -a) = a^6$
    ...

    $\therefore (-a)^n = a^n$ if n is even

    While if $n = 3, 5, 7, ...$

If $n = 3$: $(-a)^3 = (-a \times -a) \times -a = -a^3$
$(-a)^5 = (-a)^4 \times -a = -a^5$
$(-a)^7 = (-a)^6 \times -a = -a^7$

$\therefore (-a)^n = -(a^n)$
If n is odd

13. The amount of profit, x, made by a collection of medium scale business enterprises and the capital y, instead initially have the relation $y = ax^k$.
Given the corresponding values

x	4	8	12	16
y	12	34	62	96

Plot $\log_{10} y$ against $\log_{10} x$ and estimate the values of the constants a and k

x	4	8	12	16
$\log_{10} x$	0.6021	0.9031	1.0792	1.2041
y	12	34	62	96
$\log_{10} y$	1.0792	1.5315	1.7924	1.9823

Graph: PLS CHECK

$y = ax^k$

$\log_{10} y = \log_{10} ax^k$

$\log_{10} y = \log_{10} a + \log_{10} x^k$

$\log_{10} y = \log_{10} a + k\log_{10} x$
$1.0792 = \log_2 a + k \times 0.6021$ (1)
$1.5315 = \log_{10} a + k \times 0.9031$ (2)

$0.6021k + \log_{10} a = 1.0792$ (3)
$0.9031k + \log_{10} a = 1.5315$ (4)

(4) $\div$ (3)
$\dfrac{0.3010k}{0.3010} = \dfrac{0.4523}{0.3010}$

$k = 1.5026578$
$k \cong 1.5$

In equ. (4)
$\log_{10} a = 1.5315 - (0.9031 \times 1.5026578)$

$\log_{10} a = 1.5315 - 1.3570502$

$\log_{10} a = 0.1745$
$a = 10^{0.1745}$

$a = 1.493$

$a \cong 1.5$

14. Show that
   i. $(a - b)^y \neq a^y - b^y$
   ii. $a^y + a^x \neq a^{y+x}$
   iii. $b^x - b^y \neq b^{x-y}$

i. $(a-b)^y = a^y - b^y$

Proof: $(a-b) \times (a-b) \times \cdots$

$= a^y + \cdots - b^y$ (if y is an odd number)

OR

$= a^y + \cdots + b^y$ (if y is an even number)

$\therefore (a-b)^y \neq a^y - b^y$

ii. $a^y + a^x \neq a^{y+x}$

$a^{y+x} = a^y \times a^x$

$\therefore a^y + a^x \neq a^{y+x}$

iii. $b^x - b^y \neq b^{x-y}$

$b^{x-y} = \dfrac{b^x}{b^y}$

$\therefore b^x - b^y \neq b^{x-y}$

15. Using the laws of indices and logarithms, calculate

i. $31^3 \times 72$
ii. $73^2 \div 39$
iii. $64^{\frac{1}{4}} \times 29^{\frac{1}{3}}$
iv. $\dfrac{69 \times 43^3 \times \sqrt{13}}{143 \times 17}$
v. $0.09 \times 0.6 \times 1.4$

i. $31^3 \times 72$

No	Log	Calculation
31^3	3×1.4914	4.4742
$\times 72$	1.8573	+ 1.8573
		6.3315

2,145,000

Actual: 2,144.952

ii. $73^2 \div 39$

No	Log	Calculation
73^2	2×1.8633	3.7266
$\times 39$	1.5911	− 1.5911
		2.1355

$Actual: 136.6410256 \cong 136.6$

iii. $64^{\frac{1}{4}} \times 29^{\frac{1}{3}}$

No	Log	Calculation
$64^{\frac{1}{4}}$	$\frac{1}{4} \times 1.8062$	0.4516
$\times 29^{\frac{1}{3}}$	$\frac{1}{3} \times 1.4624$	+ 0.4875
		0.9391

8.692

$Actual: 8.689824246 \cong 8.690$

iv. $\dfrac{69 \times 43^3 \times \sqrt{13}}{143 \times 17}$

No	Log	Calculation
(69	1.8388	(1.8388
$\times 43^3$	$+ 1.6335 \times 3$	+ 4.9005
$\times \sqrt{13})$	$+ 1.1139 \times \frac{1}{2})$	+ 0.5570
− (143	− (2.1553)	7.2963
× 17)	+ 1.2304)	
	− 3.3857	− 3.3857
		3.9106

8,139

$Actual: 8,136,566435 \cong 8,137$

v. $0.09 \times 0.6 \times 1.4$

No	Log
0.09	$\bar{2}.9542$
× 0.6	$+ \bar{1}.7782$

× 1.4	+ 1.1461
0.7560	$\bar{1}.8785$

Actual: 0.756

16. The amount A naira to which sum of P naira accumulates in n years at r percent per annum compound interest is given by the formula

$$(A = P\left(1 + \frac{r}{100}\right)^n$$

Calculate

a. The values of r if a sum of money doubles itself in 10 years
b. The least whole number of years in which a sum of money will have trebled itself if $r = 5$.

a. $A = P\left(1 + \frac{r}{100}\right)^n$

& $A = 2P$

i.e. $2 \times P$ when $n = 10$

$\therefore$

$$\frac{2P}{P} = \frac{P\left(1 + \frac{r}{100}\right)^{10}}{P}$$

$$2 = \left(\frac{100 + r}{100}\right)^{10}$$

Add $\log_{10}$ to both sides

$$\log_{10} 2 = \log_{10} \left(\frac{100 + r}{100}\right)^{10}$$

$0.3010 = 10[\log_{10}(100 + r) - \log_{10} 100]$

$0.3010 = 10[\log_{10} 100 + r - 2]$

$20 + 0.3010 = 10\log_{10} 100 + r$

$$\frac{20.3010}{10} = \frac{10\log_{10} 100 + r}{10}$$

$2.0301 = \log_{10} 100 + r$

$10^{2.0301} = 100 + r$

$107.2 = 100 + r$

$r = 107.2 - 100$
$r = 7.2$

b. $r = 5$, $A = 3 \times P$ find n?

$$A = P\left(1 + \frac{r}{100}\right)^n$$

$$3P = P\left(1 + \frac{5}{100}\right)^n$$

$$\frac{3P}{P} = \frac{P\left(1 + \frac{1}{20}\right)^n}{P}$$

$$3 = \left(\frac{20 + 1}{20}\right)^n$$

$$3 = \left(\frac{21}{20}\right)^n$$

Add $\log_{10}$ to both sides

$$\log_{10} 3 = \log_{10}\left(\frac{21}{20}\right)^n$$

$$0.4771 = n(\log_{10} 21 - \log_{10} 20)$$
$$0.4771 = n(1.3222 - 1.3010)$$

$$\frac{0.4771}{0.0212} = \frac{n(0.0212)}{0.0212}$$

$$n \cong 22.5$$

17. Prove the identity:

$$\log_{\frac{a}{b}} x = \frac{\log_a x \log_b x}{\log_b x - \log_a x}$$

(Theorem 13)

$$\log_{\frac{a}{b}} x = \frac{1}{\log_x \frac{a}{b}}$$

$$\log_{\frac{a}{b}} x = \frac{1}{\log_x a - \log_x b}$$

$$\log_{\frac{a}{b}} x = \frac{1}{\frac{\log_a a}{\log_a x} - \frac{\log_b b}{\log_b x}}$$

$$\log_{\frac{a}{b}} x = 1 \div \left[\frac{1}{\log_a x} - \frac{1}{\log_b x}\right]$$

$$\log_{\frac{a}{b}} x = 1 \div \left[\frac{\log_b x - \log_a x}{\log_a x \log_b x}\right]$$

$$\log_{\frac{a}{b}} x = 1 \times \frac{\log_a x \log_b x}{\log_b x - \log_a x}$$

$$\therefore \log_{\frac{a}{b}} x = \frac{\log_a x \log_b x}{\log_b x - \log_a x}$$

18. Evaluate

$$81^{\frac{1}{\log_5 9}} + 3^{\frac{3}{\log_{\sqrt[3]{6}} 3}} \left[(\sqrt{7})^{\frac{2}{\log_{25} 7}} - 125^{\log_{25} 6}\right]$$

Let $81^{\frac{1}{\log_5 9}} = x$; $3^{\frac{3}{\log_{\sqrt{6}} 3}} = y$; $(\sqrt{7})^{\frac{2}{\log_{25} 7}} = z$ and $125^{\log_{25} 6} = w$

$x + y(z - w)$ (1)

$81^{\frac{1}{\log_5 9}} = 81^{\log_9 5} = x$

$9^{2\log_9 5} = x$

Add $\log_9$ to both sides

$\log_9 9^{2\log_9 5} = \log_9 x$

$2\log_9 5 \times \log_9 9 = \log_9 x$

$2\log_9 5 = \log_9 x;\ \log_9 5^2 = \log_9 x$

$\log_9 25 = \log_9 x$

$\therefore x = 25$

$3^{\frac{3}{\log_{\sqrt{6}} 3}} = y$

$3^{3\log_3 \sqrt{6}} = y$

Add $\log_3$ to both sides

$\log_3 3^{3\log_3 \sqrt{6}} = \log_3 y$

$3\log_3 \sqrt{6} \times \log_3 3 = \log_3 y$

$3\log_3 \sqrt{6} = \log_3 y$

$3\log_3 6^{\frac{1}{2}} = \log_3 y$

$\log_3 6^{\frac{3}{2}} = \log_3 y$

$\therefore y = 6^{\frac{3}{2}}$

$(\sqrt{7})^{\frac{2}{\log_{25} 7}} = z$

$7^{\frac{1}{2} \times \frac{2}{\log_{25} 7}} = z$

$$7^{\frac{1}{\log_{25} 7}} = z$$

$$7^{\log_7 25} = z$$

Add $\log_7$ to both sides

$$\log_7 7^{\log_7 25} = \log_7 z$$

$$\log_7 25 \times \log_7 7 = \log_7 z$$

$$\log_7 25 = \log_7 z$$

$$\therefore z = 25$$

$$125^{\log_{25} 6} = w$$

Add $\log_{125}$ to both sides

$$\log_{125} 125^{\log_{25} 6} = \log_{125} w$$

$$\log_{25} 6 \times \log_{125} 125 = \log_{125} w$$

$$\log_{25} 6 = \log_{125} w$$

$$\log_{5^2} 6 = \log_{5^3} w$$

$$\frac{1}{2} \log_5 6 = \frac{1}{3} \log_5 w$$

Multiply both sides by 3:

$$\frac{3}{2} \log_5 6 = \log_5 w$$

$$\log_5 6^{\frac{3}{2}} = \log_5 w$$

$$\therefore w = 6^{\frac{3}{2}}$$

Back to equation (1):

$x + y(z - w)$

Where $x = 25; y = 6^{\frac{3}{2}}; z = 25$ & $w = 6^{\frac{3}{2}}$

$25 + 6^{\frac{3}{2}}\left(25 - 6^{\frac{3}{2}}\right)$

$25 + 6^{\frac{3}{2}} \times 25 - \left(6^{\frac{3}{2}}\right)^2$

$25 + 14.69694 \times 25 - 216$

$25 + 367.4235 - 216$

$392.4235 - 216$

176.4235

Note: If the expression was:

$(x + y)(z - w)$

Knowing that $x = z$ and $y = w$

Then $(x + y)(x - y) = x^2 - y^2$

$= 25^2 - \left(6^{\frac{3}{2}}\right)^2$

$= 625 - 216 = 409$

But this isn't the expression as given in the question, therefore the former is the correct answer.

19. Find x, y such that

$$\sqrt{8 + 2\sqrt{15}} = \sqrt{x} + \sqrt{y}$$

$$8 + 2\sqrt{15} = (\sqrt{x} + \sqrt{y})^2$$

$$8 + 2\sqrt{15} = (\sqrt{x} + \sqrt{y})(\sqrt{x} + \sqrt{y})$$

$$8 + 2\sqrt{15} = x + y + 2\sqrt{xy}$$

Therefore

$x + y = 8$ & $xy = 15$

$x = 8 - y$

Then $xy = 15$

$(8 - y)y = 15$

$y^2 - 8y = -15$

$y^2 - 8y + 15 = 0$

$y^2 - 5y - 3y + 15 = 0$

$y(y - 5) - 3(y - 5) = 0$

$(y - 3)(y - 5) = 0$

$y - 3 = 0$

OR

$y - 5 = 0$

$y = 3$ or $y = 5$

If $y = 3$ then $x = 5$
If $y = 5$ then $x = 3$

20. Evaluate $\log_{0.75}\left(\log_2 \sqrt{\sqrt[-2]{0.125}}\right)$

$\log_{0.75}\left[\log_2 \sqrt{(0.125)^{-1/2}}\right]$

$\log_{0.75}\left[\log_2 \sqrt{\left(\dfrac{125}{1000}\right)^{-1/2}}\right]$

$\log_{0.75}\left[\log_2 \sqrt{\left(\dfrac{1}{8}\right)^{-1/2}}\right]$

$\log_{0.75}\left[\log_2 \sqrt{\left(1 \div \dfrac{1}{8}\right)^{1/2}}\right]$

$$\log_{0.75}\left[\log_2 \sqrt{8^{1/2}}\right]$$

$$\log_{0.75}\left[\log_2 \left(8^{1/2}\right)^{1/2}\right]$$

$$\log_{0.75}\left[\log_2 8^{1/4}\right]$$

$$\log_{0.75}\left[\frac{1}{4}\log_2 2^3\right]$$

$$\log_{0.75}\left[3 \times \frac{1}{4}\log_2 2\right]$$

$$\log_{\frac{75}{100}} \frac{3}{4}$$

$$\log_{\frac{3}{4}} \frac{3}{4}$$

$$= 1$$

21. Simplify and rationalize the denominator

$$\frac{1}{1+\sqrt{3}-\sqrt{2}} + \frac{1}{1+\sqrt{3}+\sqrt{2}}$$

$$\frac{1+\sqrt{3}+\sqrt{2}+1+\sqrt{3}-\sqrt{2}}{(1+\sqrt{3}-\sqrt{2})(1+\sqrt{3}+\sqrt{2})}$$

$$\frac{2+2\sqrt{3}}{1+\sqrt{3}+\sqrt{2}+\sqrt{3}+3+\sqrt{6}-\sqrt{2}-\sqrt{6}-2}$$

$$\frac{2+2\sqrt{3}}{2+2\sqrt{3}} = 1$$

22. Solve the equation

$$\sqrt{(3x+4)} - \sqrt{(x+2)} = \sqrt{(x-3)}$$

$$x - 3 = \left(\sqrt{3x+4} - \sqrt{x+2}\right)^2$$

$$x - 3 = (\sqrt{3x + 4} - \sqrt{x + 2})(\sqrt{3x + 4} - \sqrt{x + 2})$$

$$x - 3 = 3x + 4 - \sqrt{(3x + 4)(x + 2)} - \sqrt{(3x + 4)(x + 2)} + x + 2$$

$$x - 3 = 4x + 6 - 2\sqrt{3x^2 + 6x + 4x + 8}$$

$$x - 3 = 4x + 6 - 2\sqrt{3x^2 + 10x + 8}$$

$$\frac{2\sqrt{3x^2 + 10x + 8}}{2} = \frac{4x - x + 6 + 3}{2}$$

$$\sqrt{3x^2 + 10x + 8} = \frac{3x + 9}{2}$$

$$3x^2 + 10x + 8 = \left(\frac{3x + 9}{2}\right)^2$$

$$3x^2 + 10x + 8 = \frac{9x^2 + 27x + 27x + 81}{4}$$

$$12x^2 + 40x + 32 = 9x^2 + 54x + 81$$

$$12x^2 - 9x^2 + 40x - 54x + 32 - 81 = 0$$

$$3x^2 - 14x - 49 = 0$$

$$3x^2 - 14x - 49 = 0$$

$$3x^2 - 21x + 7x - 49 = 0$$

$$3x(x - 7) + 7(x - 7)$$

$$(3x - 7)(x - 7) = 0$$

$$3x + 7 = 0 \text{ OR } x - 7 = 0$$

$$x = -\frac{7}{3} \text{ or } x = 7$$

If $x = -\frac{7}{3}$

$$\sqrt{-7 + 4} - \sqrt{\frac{13}{3}} = \sqrt{-\frac{2}{3}}$$

Complex number involved

If $x = 7$

$$\sqrt{21 + 4} - (\sqrt{7 + 2} = \sqrt{7 - 3}$$

$$\sqrt{25} - \sqrt{9} = \sqrt{4}$$

$$5 - 3 = 2$$

$$\therefore x = 7$$

23. Rationalize $\dfrac{xy}{\sqrt{x^2+y^2}+x}$

$$\dfrac{xy}{\sqrt{x^2+y^2}+x} \times \dfrac{\sqrt{x^2+y^2}-x}{\sqrt{x^2+y^2}-x}$$

$$\dfrac{xy\left(\sqrt{x^2+y^2}-x^2y\right)}{x^2+y^2-x\sqrt{x^2+y^2}+x\sqrt{x^2+y^2}-x^2}$$

$$\dfrac{xy\sqrt{x^2+y^2}-x^2y}{y^2}$$

$$\dfrac{xy\left(\sqrt{x^2+y^2}-x\right)}{y^2}$$

$$\dfrac{x\left(\sqrt{x^2+y^2}-x\right)}{y}$$

$$= \dfrac{x}{y}\left(\sqrt{x^2-y^2}-x\right)$$

24. Simplify

$$\dfrac{2\sqrt{3}-3\sqrt{2}}{2\sqrt{3}+3\sqrt{2}}$$

$$= \dfrac{2\sqrt{3}-3\sqrt{2}}{2\sqrt{3}+3\sqrt{2}} \cdot \dfrac{2\sqrt{3}-3\sqrt{2}}{2\sqrt{3}-3\sqrt{2}}$$

$$= \dfrac{4\cdot 3 - 6\sqrt{6} - 6\sqrt{6} + 9\cdot 2}{4\cdot 3 - 6\sqrt{6} + 6\sqrt{6} - 9\cdot 2}$$

$$\dfrac{12 + 18 - 12\sqrt{6}}{12 - 18}$$

$$\dfrac{30 - 12\sqrt{6}}{-6}$$

$$\frac{6(5-2\sqrt{6})}{-6}$$

$$-(5-2\sqrt{6})$$

$$=2\sqrt{6}-5$$

25. Express in terms of simpler surds:
    a. $\sqrt{12}$
    b. $\sqrt{8}$
    c. $\sqrt{250}$
    d. $\sqrt{28}$

    a. $\sqrt{4}\times\sqrt{3}=2\sqrt{3}$
    b. $\sqrt{8}=\sqrt{4}\times\sqrt{2}=2\sqrt{2}$
    c. $\sqrt{250}=\sqrt{25}\times\sqrt{10}=5\sqrt{10}$
    d. $\sqrt{28}=\sqrt{4}\times\sqrt{7}=2\sqrt{7}$

26. Simplify $\left(\dfrac{\sqrt{2}-\sqrt{3}}{\sqrt{2}+\sqrt{3}}\right)^2$

$$\left(\frac{\sqrt{2}-\sqrt{3}}{\sqrt{2}+\sqrt{3}}\cdot\frac{\sqrt{2}-\sqrt{3}}{\sqrt{2}-\sqrt{3}}\right)$$

$$\left(\frac{2-\sqrt{6}-\sqrt{6}+3}{2-\sqrt{6}+\sqrt{6}-3}\right)^2$$

$$\left(\frac{5-2\sqrt{6}}{-1}\right)^2$$

$$(2\sqrt{6}-5)^2$$

$$(2\sqrt{6}-5)(2\sqrt{6}-5)$$

$$=4\times 6 - 10\sqrt{6} - 10\sqrt{6} + 25$$

$$24 - 20\sqrt{6} + 25$$

$$=49 - 20\sqrt{6}$$

27. Simplify $5\sqrt{75}-\sqrt{12}+\sqrt{27}$

$$5\sqrt{25 \times 3} - \sqrt{4 \times 3} + \sqrt{9 \times 3}$$

$$= 25\sqrt{3} - 2\sqrt{3} + 3\sqrt{3}$$

$$= 26\sqrt{3}$$

28. Rationalize $\dfrac{\sqrt{1-x} + \sqrt{1+x}}{\sqrt{1-x} - \sqrt{1+x}}$

$$\dfrac{\sqrt{1-x} + \sqrt{1+x}}{\sqrt{1-x} - \sqrt{1+x}} \cdot \dfrac{\sqrt{1-x} + x\sqrt{1+x}}{\sqrt{1-x} + x\sqrt{1+x}}$$

$$\dfrac{\sqrt{1-x} + \sqrt{1+x}}{\sqrt{1-x} - \sqrt{1+x}} \times \dfrac{\sqrt{1-x} + \sqrt{1+x}}{\sqrt{1-x} + \sqrt{1+x}}$$

$$\dfrac{1 - x + \sqrt{1-x^2} + \sqrt{1-x^2} + 1 + x}{1 - x + \sqrt{1-x^2} - \sqrt{1-x^2} - (1+x)}$$

$$\dfrac{2 + 2\sqrt{1-x^2}}{-2x}$$

$$\dfrac{2\left(1 + \sqrt{1+x^2}\right)}{-2x}$$

$$= -\dfrac{1}{x}\left(1 + \sqrt{1-x^2}\right)$$

29. Evaluate $\sqrt{14 + 4\sqrt{10}}$

$$\sqrt{14 + 4\sqrt{10}} = \sqrt{x} + \sqrt{y} \quad (1)$$

$$14 + 4\sqrt{10} = x + y + 2\sqrt{xy}$$

$$x + y = 14 \quad (3)$$

If this then

$$\sqrt{14 - 4\sqrt{10}} = \sqrt{x} - \sqrt{y} \quad (2)$$

$$14 - 4\sqrt{10} = (\sqrt{x} - \sqrt{y})^2$$

$14 \cdot 4\sqrt{10} = x + y - 2\sqrt{xy}$

Multiply equ. (1) and (2):

$\sqrt{14^2 - 16 \times 10} = x - y$

$\sqrt{196 - 160} = x - y$

$\sqrt{36} = x - y$

$x - y = 6$

Add (3) & (4)

$x + y = 14$

$x - y = 6$

$2x = 20$

$x = 10$

If $x = 10$ then $y = 4$

$\therefore \sqrt{14 + 4\sqrt{10}} = \sqrt{10} + \sqrt{4}$

$\sqrt{14 + 4\sqrt{10}} = 2 + \sqrt{10}$

ALGEBRAIC EQUATIONS

Algebraic Expressions
An equation is a statement that two quantities are equal e.g.
$2 + 3 = 5$

Algebra (A Noun) is a type of mathematics that uses letters and other signs to represent numbers and values.

Algebraic (adjective) equation (Noun)
Therefore: An Algebraic Equation is a statement that two quantities, which involve letters or signs, are equal. The letters or signs represent numbers or values.

Definition 1: The four operations of addition, subtraction, multiplication and division are called rational operations of arithmetic.

Algebraic operations of Arithmetic include the rational operations, the operations of raising to a power and taking roots.

Definition 2: The equation of the form $ax^2 + bx + c = 0$, where $a \neq 0$, b and c are real numbers; is called a quadratic equation in a variable x. For a particular numerical value of x, the quadratic expression $ax^2 + bx + c$ would be equal to a constant, k, say. The value of x which satisfy the equation $ax^2 + bx + c = 0$ are called the roots or solutions of the equation.

Solution by Factorization
Suppose that
$$ax^2 + bx + c \equiv (x - \alpha)(x - \beta)$$

This method of factorization gives the two solutions of the quadratic equation.
In general, suppose that there are integers, p, q, r and s such that
$$ax^2 + bx + c \equiv (px + q)(rx + s)$$

Note: Such factorization as the one above is not always possible.

Then
$$ax^2 + bx + c \equiv prx^2 + psx + qrx + qs$$

$$ax^2 + bx + c \equiv prx^2 + (ps + qr)x + qs$$

So that
$a = pr$
$b = ps + qr$
$c = qs$

If $a = 1$, then $p = 1$ and $r = 1$ (or $p = -1$ & $r = -1$)

If $p = 1$ and $r = 1$, then $b = s + q$ and $c = qs$

In order to obtain the required $q, s,;$

First (i) write down all possible pairs (q,s) such that $q \times s = c$,

Then
(ii) examine which pair (if any) satisfies $b = s + q$.

If such a pair exists, we are home.

If not, no factors (real numbers) of the given form exists.

If $a \neq 1$, first
i. Express $a \times c$ as a product of two factors in such a way as to
ii. Break bx into 2 terms, so that
iii. The quadratic function is written with four terms which can be factorized by the grouping method.

Note: Always check that the solutions to a quadratic equation satisfy the equation itself.

Proposition: The method of factorization described above can be used only if the expression:

$b^2 - 4ac$

Which is called the 'Discriminant' of the Quadratic expression

$ax^2 + bx + c$, is a perfect square i.e. Factorization works only if the Discriminant of the Quadratic Equation is a perfect square.

Solution by Completing the Square

Suppose that the discriminant, $b^2 - 4ac$, is not a perfect square. Then factorization into linear factors with rational coefficients is not possible. We need a more general method. One such method is by 'Completing the square'. [we want to believe that such equations which have the discriminant not a perfect square, have an incomplete square. So we solve the equation by adding a possible number that completes the square.]

Given the quadratic equation

$ax^2 + bx + c = 0, a \neq 0$ divide through by a to obtain $x^2 + \frac{b}{a}x + \frac{c}{a} = 0 \quad x^2 + \frac{b}{a}x = -\frac{c}{a}$.

Now complete the square part of which is $x^2 + \frac{b}{a}x$ (RHS)

Since $(x + \alpha)^2 = x^2 + 2\alpha x + \alpha^2$, the relationship between $2\alpha x + \alpha^2$ in the equation is

(i) The 2 and x was removed:

$$\frac{2ax}{2x}$$

(ii) The a was doubled to give a^2
∴ we must have
$$x^2 + \frac{b}{a}x + \left(\frac{b}{2a}\right)^2$$

[i.e. $\frac{b}{a}x$

(i) Double answer i.e. $\left(\frac{b}{2a}\right)^2$.
So we have
$$x^2 + \frac{b}{a}x + \left(\frac{b}{2a}\right)^2$$

As the square we need. But since we are considering the LHS of an equation, completing the square involves adding $\left(\frac{b}{2a}\right)^2$. We must now also add $\left(\frac{b}{2a}\right)^2$ to $\frac{c}{a}$ in order to leave our equation balanced:

Adding $\left(\frac{b}{2a}\right)^2$ to both sides.

$$x^2 + \frac{b}{a}x + \left(\frac{b}{2a}\right)^2 = -\frac{c}{a} + \left(\frac{b}{2a}\right)^2$$

i.e.
$$x^2 + \frac{b}{a}x + \left(\frac{b}{2a}\right)^2 = \left(\frac{b}{2a}\right)^2 - \frac{c}{a}$$

$$\left(x + \frac{b}{2a}\right)^2 = \frac{b^2}{4a^2} - \frac{c}{a}$$

$$\left(x + \frac{b}{2a}\right)^2 = \frac{b^2 - 4ac}{4a^2}$$

If $b^2 - 4ac > 0$, the roots are real and distinct.

If $b^2 - 4ac = 0$, the roots are real and equal.

If $b^2 - 4ac < 0$, there are no real roots. (i.e. No root satisfies the equation)

Symmetric Properties of the Roots
Let α and β denote the roots (whenever they exist) of the quadratic equation

$ax^2 + bx + c = 0$
$a \neq 0$

Check, using the formula, that
$\alpha + \beta = -\frac{b}{a}$, $\alpha\beta = \frac{c}{a}$ (Valid)

Note (1): The formula for the sum and product of roots given above can be used to check whether the solutions of a quadratic equation are correct or not.

Note (2): No 5 of Exercise 3.2 shows important symmetric identities.

Graphical Methods of Solution
The graph of a polynomial equation $p(x,y) = 0$ is the set of points in the plane the ordered coordinates of which are member of the solution set $p(x,y) = 0$.

Definition 2: The graph of $y = ax^2 + bx + c$, $a \neq 0$ is called a **parabola**. Its highest or lowest point is called **vertex**.

Properties of the Parabola.
Given the parabola with equation
$y = ax^2 + bx + c$, $a \neq 0$.

(i) If $a > 0$, the graph opens upward;
(ii) If $a < 0$, the graph opens downwards;
(iii) If $ax^2 + bx + c = 0$ has real solutions or roots x_1 and x_2, where $x_1 \neq x_2$ (or if the discriminant is greater then 0, $b^2 - 4ac > 0$), the graph crosses the x-axis at $x = x_1$ and $x = x_2$;

GRAPH PLS CHECK

(iv) If $ax^2 + bx + c = 0$ has two equal roots, the graph touches the x-axis at the point $x = x_1 = x_2$. This means that the point x-axis is tangent to the graph at $x = x_1 + x_2$; (i.e. where $b^2 - 4ac = 0$)

(v) If $ax^2 + bx + c = 0$ has no roots, it is said to have imaginary roots, the graph neither crosses nor touches the x-axis; (i.e. where $b^2 - 4ac < 0$)

GRAPH PLS CHECK

(vi) The graph is symmetric about the line $x = -\dfrac{b}{2a}$;

(vii) The vertex of the parabola is the point $\left(-\dfrac{b}{2a}, \dfrac{-b^2 + 4ac}{4a}\right)$;

(viii) Finally, when $a = 0$, the graph is a straight line $y = bx + c$, which is completely determined by two points; or

(ix) By the gradient b and intercept c.

Suppose that instead of $y = ax^2 + bx + c$, we have $ax^2 + bx + c = k$ or $ax^2 + bx + c = mx + k$.

Then, instead of intersections with the x-axis (i.e. the line $y = 0$) the solutions we now seek are the x-coordinates of the points of intersection of the graphs of the curve

$y = ax^2 + bx + c$

And the line $y = k$ or the line $y = mx + k$.

Solution of two Simultaneous Equation where one is linear and the other is Quadratic

Practical Applications in Word Problems.

Exercise 3.1

1. Show that the equation:
   $x^2 + 3x - 10 = 0$ has roots 2 and -5

   By Factorization
   Discriminant: $b^2 - 4ac$ where $a = 1, b = 3$ & $c = -10$;

   $3^2 - 4(1 \times -10)$
   $9 + 40 = 49 = 7^2$

   $\therefore x^2 + 3x - 10 = 0$

   $x^2 + 5x - 2x - 10 = 0$

   $x(x + 5) - 2(x + 5) = 0$

   $(x - 2)(x + 5) = 0$

   $(x - 2) = 0$ or $(x + 5) = 0$

   $\therefore 2$ & -5 are the roots of the equation

2. Solve the quadratic equation $x^2 + 3x + 4 = 0$

   By factorization
   Discriminant: $b^2 - 4a$, where $a = 1, b = 3$ & $c = 4$;

   $3^2 - 4(1 \times 4)$

   $9 - 16 = -5$

   Equation can't be solved by factorization

3. By rewriting the following equation without log, show that $x = 6$ and -1 satisfy the equation
   $\log_{10}(x^2 - 5x + 94) = 2$

   $x^2 - 5x + 94 = 10^2$

$$x^2 - 5x + 94 - 10^2 = 0$$

$$x^2 - 5x + 94 - 100 = 0$$

$$x^2 - 5x - 6 = 0$$

$$x^2 - 6x + x - 6 = 0$$

$$x(x - 6) + 1(x - 6) = 0$$

$$(x + 1)(x - 6) = 0$$

$$(x + 1) = 0 \text{ OR } (x - 6) = 0$$

$$x = -1 \ \& \ x = 6$$

Therefore $6 \ \& \ -1$ satisfy the equation

Exercise 3.2

1. Solve the quadratic equation
   $15t^2 - 4t + 2 = 0$

   Discriminant: $(-4)^2 - 4(15 \times 2)$

   $16 - 20 = -104$ which is less then 0

   $\therefore$ There are no real roots for the equation

2. Solve the equation $x^2 + 4x + 16 = 0$
   Discriminant: $(4)^2 - 4(1 \times 16)$

   $16 - 64 = -48 < 0$ (No real roots)

3. Solve the equation $4x^2 + 3x + 1 = 0$

   Discriminant $(3)^2 - 4(4 \times 1)$

   $9 - 16 = -5 < 0$ (No real roots)

4. Solve the equation
   $t^2 + 3t - 5 = 0$

   Discriminant

   $3^2 - 4(1 \times -5); \ 9 + 20 = 29 > 0$ (distinct real roots)

   By formula $t = \dfrac{-b \pm \sqrt{b^2 - 4ac}}{2a}$

$a = 1, b = 3$ & $c = -5$

$$t = \frac{-3 \pm \sqrt{3^2 - 4(1 \times -5)}}{2(1)}$$

$$t = \frac{-3 \pm \sqrt{29}}{2}$$

$$\therefore t = \frac{-3 + \sqrt{29}}{2} \quad \text{or} \quad t = \frac{-3 - \sqrt{29}}{2}$$

5. Given $\alpha + \beta$ and $\alpha\beta$, prove that

   i. $\alpha^2 + \beta^2 = (\alpha + \beta)^2 - 2\alpha\beta$

   $\alpha^2 + \beta^2 = (\alpha + \beta)^2 - 2\alpha\beta$ (1)

   $(\alpha + \beta)^2 - 2\alpha\beta = \alpha^2 + 2\alpha\beta + \beta^2 - 2\alpha\beta$

   $\alpha^2 + \beta^2 + 2\alpha\beta - 2\alpha\beta = \alpha^2 + \beta^2$

   Equation (1) is valid.

   ii. $\alpha^3 + \beta^3 = (\alpha + \beta)^3 - 3\alpha\beta(\alpha + \beta)$ (2)

   $(\alpha + \beta)^3 - 3\alpha\beta(\alpha + \beta) = (\alpha + \beta)(\alpha^2 + 2\alpha\beta + \beta^2) - 3\alpha\beta(\alpha + \beta)$

   $\alpha^3 + 2\alpha^2\beta + \alpha\beta^2 + \alpha^2\beta + 2\alpha\beta^2 + \beta^3 - 3\alpha\beta(\alpha + \beta)$

   $\alpha^3 + \beta^3 + 3\alpha^2\beta + 3\alpha\beta^2 - 3\alpha\beta(\alpha + \beta)$

   $\alpha^3 + \beta^3 + 3\alpha^2\beta + 3\alpha\beta^2 - 3\alpha\beta(\alpha + \beta)$

   $\alpha^3 + \beta^3 + 3\alpha^2\beta + 3\alpha\beta^2 - 3\alpha\beta(\alpha + \beta)$

   $\alpha^3 + \beta^3 + 3\alpha^2\beta + 3\alpha\beta^2 - 3\alpha^2\beta - 3\alpha\beta^2 = \alpha^3 + \beta^3$

   Equation (2) is valid

   iii. $(\alpha - \beta)^2 - (\alpha + \beta)^2 - 4\alpha\beta$ (3)

   $(\alpha + \beta)^2 - 4\alpha\beta$

   $\alpha^2 + 2\alpha\beta + \beta^2 - 4\alpha\beta = \alpha^2 + \beta^2 - 2\alpha\beta$

   &

   $(\alpha - \beta)^2 = (\alpha^2 + 2\alpha\beta + \beta^2)$

   $\therefore$ Equation (3) is valid

iv. $\dfrac{1}{\alpha}+\dfrac{1}{\beta}=\dfrac{\alpha+\beta}{\alpha\beta}$

$\dfrac{1}{\alpha}+\dfrac{1}{\beta}=\dfrac{\beta+\alpha}{\alpha\beta}$

$\dfrac{\alpha+\beta}{\alpha\beta}$

Equation (4) is valid

v. $\alpha^2\beta^2 = (\alpha\beta)^2$ (5)

Law of indices

Equation (5) is valid

vi. $\alpha^4 + \beta^4 = (\alpha^2 + \beta^2)^2 - 2\alpha^2\beta^2$ (6)

$(\alpha^2 + \beta^2)^2 - 2\alpha^2\beta^2$

$\alpha^4 + 2\alpha^2\beta^2 + \beta^4 - 2\alpha^2\beta^2$

$\alpha^4 + \beta^4 + 2\alpha^2\beta^2 - 2\alpha^2\beta^2 = \alpha^4 + \beta^4$

Equation (6) is valid

& $(\alpha^2 + \beta^2)^2 - 2\alpha^2\beta = [(\alpha + \beta)^2 - 2\alpha\beta]^2 - 2(\alpha\beta)^2$

Exercise 3.3

1. Let a, b and c denote real constants. Show that if the quadratic equation $x^2 - (3c - b)x + bc = 0$ has equal roots, then so does $x^2 - (5c - b)x + 4x^2 = 0$.

    Discriminant of $x^2 - (3c - b)x + bc = 0$

    $[-(3c - b)]^2 - 4(1 \times bc)$

    $= 9c^2 - 6bc + b^2 - 4bc$

$9c^2 - 10bc + b^2;$

$0 = 9c^2 - 10bc + b^2$

Discriminant of $x^2 - (5c - b)x + 4c^2 = 0$

$[-(5c-b)]^2 - 4(1 \times 4c^2)$

$25c^2 - 10bc + b^2 - 16c^2$

$9c^2 - 10bc + b^2;$

$0 = 9c^2 - 10bc + b^2$

$9c^2 + 10bc + b^2 = 0$

$9c^2 - 9bc - bc + b^2 = 0$

$9c^2 - 9bc - bc + b^2 = 0$

$9c(c-b) - b(c-b) = 0$

$(9c-b)(c-b) = 0$

$9c = b$ or $c = b$

If $9c = b$

$9c^2 - 10(9c^2) + 81c^2$

$90c^2 - 90c^2 = 0$

And if $c = b$

$9b^2 - 10b^2 + b^2 = 0$

$10b^2 - 10b^2 = 0$

Since both discriminant are equal then both have equal roots for each.

2. If α and β are the roots of the equation

$2x^2 + 7x + 3 = 0$

Obtain the equation the roots of which are $\dfrac{1}{\alpha^2}$ and $\dfrac{1}{\beta^2}$.

$\alpha + \beta = -\dfrac{7}{2}$ & $\alpha\beta = \dfrac{3}{2}$ (roots of $2x^2 + 7x + ...$)

$\dfrac{1}{\alpha^2} + \dfrac{1}{\beta^2} = \dfrac{\alpha^2 + \beta^2}{\alpha^2\beta^2}$

$\dfrac{\alpha^2 + \beta^2}{\alpha^2\beta^2} = \dfrac{(\alpha + \beta)^2 - 2\alpha\beta}{(\alpha\beta)^2}$

When $\alpha + \beta = -\dfrac{7}{2}$ & $\alpha\beta = \dfrac{3}{2}$

Then $\dfrac{\left(-\dfrac{7}{2}\right)^2 - 2\left(\dfrac{3}{2}\right)}{\left(\dfrac{3}{2}\right)^2} = \dfrac{\dfrac{49}{4} - \dfrac{6}{2}}{\dfrac{9}{4}}$

$\left(\dfrac{49 - 12}{4}\right) \times \dfrac{4}{9} = \dfrac{37}{9}$

$\dfrac{1}{\alpha^2} \times \dfrac{1}{\beta^2} = \dfrac{1}{\alpha^2\beta^2} = \dfrac{1}{(\alpha\beta)^2}$

$\dfrac{1}{(\alpha\beta)^2} = \dfrac{1}{\left(\dfrac{3}{2}\right)^2} = 1 \div \dfrac{9}{4} = \dfrac{4}{9}$

∴ the equation being that

$\dfrac{1}{\alpha^2} + \dfrac{1}{\beta^2} = \dfrac{37}{9}$

&
$\alpha^2\beta^2 = \dfrac{4}{9}$

The equation is $x^2 - \left(\dfrac{37}{9}\right)x + \dfrac{4}{9}$

$x^2 - \dfrac{37}{9}x + \dfrac{4}{9}$

$9x^2 - 37x + 4 = 0$

Exercise 3.4

1. Solve the equation
   $x^2 - 3x + 5 = 0$ by
   i. drawing the graph of $f(x)$ and noting the intersection on the x-axis
   ii. setting $f(x)$ so that $x^2 = 3x - 5$ on the same scale.
      Obtain the x- coordinates of the intersection as the solutions. Compare the solution (i) and (ii)

   $f(x) = x^2 - 3x + 5$

   $f(x) = 0$

   From the question find the coordinates of x that equate the equation to 0 i.e. $f(x) = 0$

i. Table of Values

x	-3	-2	-1	0	1	2	3	4
x^2	9	4	1	0	1	4	9	16
$-3x$	9	6	3	0	-3	-6	-9	-12
5	5	5	5	5	5	5	5	5
$y/f(x)$	23	15	9	5	3	3	5	9

Check discriminant: $b^2 - 4ac$

$(-3)^2 - 4(1 \times 5); 9 - 20 = -11$ (No real roots)

GRAPH

ii.

x	-3	-2	-1	0	1	2	3	4
x^2	9	4	1	0	1	4	9	16
$-3x$	9	6	3	0	-3	-6	-9	-12
$3x - 5$	-14	-11	-8	-5	-2	1	4	7

GRAPH

There are no intersection in both graph which means the equation has not real roots.

2. Solve the equation $x^2 - 4x - 5 = 0$
   Graphically

x	-2	-1	0	1	2	3	4	5	6
x^2	4	1	0	1	4	9	16	25	36
$-4x$	8	4	0	-4	-8	-12	-16	-20	-24

| $x^2 - 4x - 5$ | 7 | 0 | -5 | -8 | -9 | -8 | -5 | 0 | 7 |

GRAPH

The roots are -1 & 5

3. By drawing the graphs of $-3x^2$ and $4x + 2$ on the same scale, and reading off the x-coordinates of their intersection, solve the equation, $3x^2 + 4x + 2 = 0$

x	-3	-2	-1	=	1	2	3
x^2	-27	-12	-3	0	-3	-12	-27
$4x + 2$	-10	-6	-2	2	6	10	14

GRAPH

There are no real roots for the equation $3x^2 + 4x + 2 = 0$

Exercise 3.5

1. Solve the simultaneous equation
   $y = 5 - x$ (1)
   $x^2 - 2y^2 - 1 = 0$ (2)

   Since $y = 5 - x$

   Substitute in (2)

   $x^2 - 2(5 - x)^2 - 1 = 0$
   $x^2 - 2(25 - 5x - 5x + x^2) - 1 = 0$

   $x^2 - 2(25 - 5x - 5x + x^2) - 1 = 0$

   $x^2 - 2x^2 + 20x - 50 - 1 = 0$

   $-x^2 + 20x - 51 = 0$

   $x^2 - 20x + 51 = 0$

   $x^2 - 17x - 3x + 51 = 0$

   $x(x - 17) - 3(x - 17) = 0$

   $(x - 3)(x - 17) = 0$

   $x - 3 = 0$ or $x - 17 = 0$

   $x = 3$ or $x = 17$

If $x = 3$

Substitute x in (1)

$y = 5 - x$; $y = 5 - 3$

$y = 2$

If $x = 17$

Substitute x in (2)

$y = 5 - x$; $y = 5 - 17$

$y = -12$

$\therefore y = -12$, $x = 17$; or

$y = 2, x = 3$

Exercise 3.6
1. The ratio of the ages of two brothers, Sola and Femi, is 3:4. In Six years' time, the ratio of their ages will be 5:6. How old are they now?

Let their ages be x & y respectively.

$\dfrac{x}{y} = \dfrac{3}{4}$ (1)

$\dfrac{x+6}{y+6} = \dfrac{5}{6}$ (2)

$x = \dfrac{3y}{4}$ substitute x in $\left[\dfrac{3y}{4} + \dfrac{6}{1}\right] \div y + 6 = \dfrac{5}{6}$

$\dfrac{3y + 24}{4} \times \dfrac{1}{y+6} = \dfrac{5}{6}$

$\dfrac{3y + 24}{4y + 24} = \dfrac{5}{6}$

$18y + (24 \times 6) = 20y + (24 \times 5)$

$(24 \times 6) - (24 \times 5) = 2y$

$24 = 2y$ ∴ $y = 12$

If $y = 12$ then $x = \dfrac{3 \times 12}{4}$

$x = 9$ ∴ $y = 12$

i.e. Sola is 9 and Femi is 12

2. The sum of the roots of a quadratic equation is $-\dfrac{7}{2}$ and the product of its roots is $\dfrac{3}{2}$. Find the roots

Let the roots be α and β

$\alpha + \beta = -\dfrac{7}{2}$; (1)

$\alpha\beta = \dfrac{3}{2}$ (2)

In equation (1)
$\alpha + \beta = -\dfrac{7}{2}$

$\alpha = -\dfrac{7}{2} - \beta$

Substitute α in (2)

$\left(-\dfrac{7}{2} - \dfrac{\beta}{1}\right)\beta = \dfrac{3}{2}$;

$-\dfrac{\beta^2}{1} - \dfrac{7\beta}{2} = \dfrac{3}{2}$

$-\dfrac{\beta^2}{1} - \dfrac{7\beta}{2} = \dfrac{3}{2}$

$\dfrac{\beta^2}{1} + \dfrac{7\beta}{2} + \dfrac{3}{2} = 0$

Multiply both sides by 2.

$2\beta^2 + 7\beta + 3 = 0$

$2\beta^2 + 6\beta + \beta + 3 = 0$

$2\beta(\beta + 3) + 1(\beta + 3) = 0$

$(2\beta + 1)(\beta + 3) = 0$

$2\beta + 1 = 0; \quad \beta + 3 = 0$

$\beta = -\dfrac{1}{2}$ OR $\beta = -3$

If $\beta = -1/2$ in equation (1)

$\alpha + \beta = -\dfrac{7}{2}$

$\alpha = -\dfrac{7}{2} - \dfrac{\beta}{1}$

$\alpha = -\dfrac{7}{2} - \left(-\dfrac{1}{2}\right)$

$\alpha = -\dfrac{7}{2} + \dfrac{1}{2}$

$\alpha = -\dfrac{6}{2}; \quad \alpha = -3$

If $\beta = -3$

$\alpha = -\dfrac{7}{2} - (-3)$

$\alpha = -\dfrac{7}{2} + \dfrac{3}{1}$

$\alpha = \dfrac{-7 + 6}{2}$

$\alpha = -\dfrac{1}{2}$

∴ the roots are

$-\dfrac{1}{2}$ & -3

3. A motorist covers a distance of $100km$ at a certain speed. If he increases his speed by $100 km \cdot hr^{-1}$ he takes $\dfrac{1}{2}$ hour less time. Find his original speed.

Let the certain speed be $x \text{ km} \cdot hr^{-1}$

$$Time = \frac{100}{x} hr$$

$$\frac{100}{x+10} = \frac{100}{x} - \frac{1}{2}$$

$$\frac{100}{x+10} = \frac{200-x}{2x}$$

$$200x = (200-x)(x+10)$$

$$200x = 200x - x^2 + 2000 - 10x$$

$$200x - 200x = -x^2 - 10x + 2000$$

$$x^2 + 10x - 2000 = 0$$

$$x^2 + 50x - 40x - 2000 = 0$$

$$x(x+50) - 40(x+50) = 0$$

$$(x-40)(x+50) = 0$$

$$x = 40 \text{ Or } x = -50$$

Since speed is not a negative value

$$\therefore \text{ Speed} = 40 \text{ km} \cdot hr^{-1}$$

4. An object is projected vertically upwards so that its height in metres above the ground after 5 seconds is $280s - 49s^2$. After what time is it 336 metres above the ground?

$$280s - 49s^2 = 336$$

$$49s^2 - 280s + 336 = 0$$

Divide both sides by 7

$$\frac{49s^2}{7} - \frac{280}{7}s + \frac{336}{7} = \frac{0}{7}$$

$$7s^2 - 40s + 48 = 0$$

Using formula

$$\frac{-b \pm \sqrt{b^2 - 4ac}}{2a}$$

Where $a = 7$, $b = -40$ & $c = 48$

$$\frac{-(-40) \pm \sqrt{(-40)^2 - 4(7 \times 48)}}{2 \times 7}$$

$$\frac{40 \pm \sqrt{1600 - 4(338)}}{14}$$

$$= \frac{40 \pm \sqrt{1600 - 1344}}{14}$$

$$\frac{40 \pm \sqrt{256}}{14}$$

$$= \frac{40 + 16}{14} \text{ or } \frac{40 - 16}{14}$$

$$\frac{56}{14} \text{ or } \frac{24}{14}$$

4 or $\frac{12}{7}$

$\therefore$ After 4 or 12/7 seconds

5. Solve the simultaneous equation

$$y = ax^2 + bx + c$$

$$y = b_1 x + c_1$$

$$\therefore ax^2 + bx + c = b_1 x + c_1$$

$$ax^2 + bx - b_1 x + c - c_1 = 0$$

$$ax^2 + (b - b_1)x + c - c_1 = 0$$

$$x = \frac{-b \pm \sqrt{b^2 - 4ac}}{2a}$$

$$a = a, \ b = b - b_1 \ \& \ c = c - c_1$$

$$x = \frac{-(b - b_1) \pm \sqrt{(b - b_1)^2 - 4[a(c - c_1)]}}{2a}$$

Exercise 3.7

Solve by Factorization

1. $x^2 + 2x + 1 = 0$

Discriminant: $4 - 4 = 0$

$x^2 + x + x + 1 = 0$

$x(x + 1) + 1(x + 1) = 0$

$x + 1 = 0$ OR $x + 1 = 0$

$x = -1$ twice

2. $x^2 - x - 6 = 0$

Discriminant: $1 + 24 = 25$

$x^2 - 3x + 2x - 6 = 0$

$x(x - 3) + 2(x - 3) = 0$

$(x + 2)(x - 3) = 0$

$(x + 2)(x - 3) = 0$

$x = -2 \ \& \ x = 3$

3. $2x^2 + 7x + 6 = 0$

$49 - 48 = 1$

$2x^2 + 3x + 4x + 6 = 0$

$x(2x + 3) + 2(2x + 3) = 0$

$(x + 2)(2x + 3) = 0$

$x = -2$ OR $x = -\dfrac{3}{2}$

4. $5x^2 + 9x + 4 = 0$

   Discriminant: $81 - 80 = 1$

   $5x^2 + 5x + 4x + 4 = 0$

   $5x(x + 1) + 4(x + 1) = 0$

   $(5x + 4)(x + 1) = 0$

$x = -\dfrac{4}{5}$ or $x = -1$

5. $3x^2 - 7x - 6 = 0$

   Discriminant: $49 + 72 = 121$

   $3x^2 - 9x + 2x - 6 = 0$

   $3x(x - 3) + 2(x - 3) = 0$

   $(3x + 2)(x - 3) = 0$

   $x = -\dfrac{2}{3}$ or $x = 3$

6. $8x^2 + 2x - 15 = 0$

   Discriminant: $4 + 480 = 484$

   $8x^2 + 12x - 10x - 15 = 0$

   $4x(2x + 3) - 5(2x + 3) = 0$

$(4x - 5)(2x + 3) = 0$

$x = \dfrac{5}{4}$ or $x = -\dfrac{3}{2}$

7. $x^2 + x - 2 = 0$

   Discriminant: $1 + 8 = 9$

   $x^2 + 2x - x - 2 = 0$

   $x(x + 2) - 1(x + 2) = 0$

   $(x - 1)(x + 2) = 0$

   $x = 1$ or $x = -2$

8. $2x^2 - 3x - 9 = 0$

   Discriminant: $9 + 72 = 81$

   $2x^2 - 6x + 3x - 9 = 0$

   $2x(x - 3) + 3(x - 3) = 0$

   $(2x + 3)(x - 3) = 0$

   $x = -\dfrac{3}{2}$ or $x = 3$

9. $x^2 + 3x + 2 = 0$

   Discriminant: $9 - 8 = 1$

   $x^2 + 2x + x + 2 = 0$

   $x(x + 2) + 1(x + 2) = 0$

   $(x + 1)(x + 2) = 0$

   $x = -1$ or $x = -2$

10. $x^2 - 2x - 3 = 0$

    Discriminant: $4 + 12 = 16$

$x^2 - 3x + x - 3 = 0$

$x(x - 3) + 1(x - 3) = 0$

$(x + 1)(x - 3) = 0$

$x = -1$ or $x = 3$

11. $5^{2x} - 6 \times 5^x + 5 = 0$

    $(5^x)^2 - 6 \times 5^x + 5 = 0$

    Let $5^x = y$

    $y^2 - 6y + 5 = 0$

    Discriminant: $36 - 20 = 16$

    $y^2 - 5y - y + 5 = 0$

    $y(y - 5) - 1(y - 5) = 0$

    $(y - 1)(y - 5) = 0$

    $y = 1$ or $y = 5$

    If $y = 1$
    $5^x = 1$
    $5^x = 5^0$
    $\therefore x = 0$

    If $y = 5$

    $5^x = 5$
    $5^x = 5^1$

    $\therefore x = 1$

12. $x^2 + 4x + 4 = 0$

    Discriminant: $16 - 16 = 0$

    $x^2 + 2x + 2x + 4 = 0$

    $x(x + 2) + 2(x + 2) = 0$

    $(x + 2)(x + 2) = 0$

    $x = -2$ twice

13. $9x^2 + 12x + 4 = 0$

Discriminant: $144 - 154 = -10$

No real roots

14. $\log_{10}(x^2 + 4) = 2 + \log_{10} x - \log_{10} 20$

$\log_{10}(x^2 + 4) = \log_{10} 100 + \log_{10} x - \log_{10} 20$

$\log_{10}(x^2 + 14) = \log_{10} 100 + \log_{10} x - \log_{10} 20$

$\log_{10}(x^2 + 4) = \log_{10}\left(\dfrac{100 \times x}{20}\right)$

$\log_{10}(x^2 + 4) = \log_{10} 5x$

$\therefore x^2 + 4 = 5x;\ x^2 - 5x + 4 = 0$

Discriminant: $25 - 16 = 9$

$x^2 - 4x - x + 4 = 0$

$x(x - 4) - 1(x - 4) = 0$

$(x - 1)(x - 4) = 0$

$x = 1$ or $x = 4$

15. $4^x - 7 \times 2^x + 12 = 0$

Let $2^x = y$

$(2^x)^2 - 7 \times 2^x + 12 = 0$

$y^2 - 7y + 12 = 0$

Discriminant: $49 - 48 = 1$

$y^2 - 3y - 4y + 12 = 0$

$y(y - 3) - 4(y - 3) = 0$

$(y - 4)(y - 3) = 0$

$y = 4$ or $y = 3$

If $y = 4$

$2^x = 4$

$2^x = 2^2$

$x = 2$

If $y = 3$

$2^x = 3$

$$x = \log_2 3$$

$$\therefore x = 2 \text{ or } x = \log_2 3$$

16. $x^2 - 3x - 4 = 0$

Discriminant: $9 + 16 = 25$

$$x^2 - 4x + x - 4 = 0$$

$$x(x-4) - 1(x-4) = 0$$

$$(x-1)(x-4) = 0$$

$$x = 1 \text{ or } x = 4$$

17. $6x^2 - 5x - 4 = 0$

Discriminant: $25 + 96 = 121$

$$6x^2 - 8x + 3x - 4 = 0$$

$$2x(3x-4) + 1(3x+4) = 0$$

$$(2x+1)(3x-4) = 0$$

$$x = -\frac{1}{2} \ \& \ x = \frac{4}{3}$$

18. $x^2 - 2x - 8 = 0$

Discriminant: $4 + 32 = 36$

$$x^2 - 4x + 2x - 8 = 0$$

$$x(x-4) + 2(x-4) = 0$$

$$(x+2)(x-4) = 0$$

$$(x+2)(x-4) = 0$$

$x = -2$ or $x = 4$

19. $x^2 - 5x = 0$

    $\dfrac{x(x-5)}{x} = \dfrac{0}{x}$

    $x - 5 = 0$

    $x = 5$

    Or $\dfrac{x(x-5)}{x-5} = \dfrac{0}{x-5}$

    $x = 0$

    $\therefore x = 5$ or $x = 0$

20. $5k^2 + 4k - 1 = 0$

    Discriminant: $16 + 20 = 36$

    $5k^2 + 5k - k - 1 = 0$

    $5k(k+1) - 1(k+1) = 0$

    $(k+1)(5k-1) = 0$

    $k = -1$ & $k = \dfrac{1}{5}$

    Solve the following quadratic equations:

21. $x^2 - 6x + 9 = 0$

    Discriminant: $36 - 36 = 0$

    $x^2 - 3x - 3x + 9 = 0$

    $x(x-3) - 3(x-3) = 0$

    $(x-3)(x-3) = 0$

    $x = 3$ twice

22. $2x^2 - x - 3 = 0$

    Discriminant: $1 + 24 = 25$

    $2x^2 - 3x + 2x - 3 = 0$

    $x(2x-3) + 1(2x-3) = 0$

$(x + 1)(2x - 3) = 0$

$x = -1$ or $x = \dfrac{3}{2}$

23. $x^2 + (3 - \sqrt{2})x - 3\sqrt{2} = 0$

Solve by formula:

$x = \dfrac{-b \pm \sqrt{b^2 - 4ac}}{2a}$

$a = 1;\ b = (3 - \sqrt{2})\ \&\ c = -3\sqrt{2}$

$x = \dfrac{(-3 - \sqrt{2}) \pm \sqrt{(3 - \sqrt{2})^2 - 4(1 \times -3\sqrt{2})}}{2 \times 1}$

$x = \dfrac{(\sqrt{2} - 3) \pm \sqrt{(9 - 6\sqrt{2} + 2) + 12\sqrt{2}}}{2}$

$x = \dfrac{(\sqrt{2} - 3) \pm \sqrt{11 + 6\sqrt{2}}}{2}$

Where $\sqrt{11 + 6\sqrt{2}} = \sqrt{x} + \sqrt{y}$ \quad (1)

$\sqrt{11 - 6\sqrt{2}} = \sqrt{x} - \sqrt{y}$ \quad (2)

In equation (1)

$11 + 6\sqrt{2} = x + y + 2\sqrt{xy}$

$11 = x + y$ \quad (3)

Multiply (1) by (2)

$\sqrt{11 + 6\sqrt{2}} \times \sqrt{11 - 6\sqrt{2}}$

$(\sqrt{x} + \sqrt{y})(\sqrt{x} - \sqrt{y})$

$\sqrt{11 \times 11 + 66\sqrt{2} - 66\sqrt{2} - 36 \times 2}$

$x - y + \sqrt{xy} - \sqrt{xy}$

$\sqrt{121 - 72} = x - y$

$\sqrt{49} = x - y$

$x - y$

$11 = x + y$
$7 = x - y$
(3) + (4): $18 = 2x$
$x = 9; y = 11 - 9 = 2$

$x = 9, y = 2$

$\therefore \sqrt{11 + 6\sqrt{2}} = \sqrt{9} + \sqrt{2}$

Back to equation

$$x = \frac{(\sqrt{2} - 3) \pm (\sqrt{9} + \sqrt{2})}{2}$$

$$x = \frac{(\sqrt{2} - 3) - (\sqrt{9} + \sqrt{2})}{2}$$

$$x = \frac{(\sqrt{2} - 3) + (\sqrt{9} + \sqrt{2})}{2}$$

$$x = \frac{-\sqrt{9} - \sqrt{2} + \sqrt{2} - 3}{2}$$

Or

$$x = \frac{\sqrt{9} + \sqrt{2} + \sqrt{2} - 3}{2}$$

$x = \dfrac{-3 - 3}{2}$ or $x = \dfrac{3 + 2\sqrt{2} - 3}{2}$

$x = -\dfrac{6}{2}$ or $x = \dfrac{2\sqrt{2}}{2}$

$x = -3$ or $x = \sqrt{2}$

24. $t^2 + t - 12 = 0$

$1 + 48 = 49$

$t^2 - 4t + 3t - 12 = 0$

$t(t - 4) + 3(t - 4) = 0$

$t = -3$ or $t = 4$

25. Prove that the roots of the equation

    $ax^2 - (a+c)x + c = 0$ are real

    $b^2 - 4ac$

    $b = -(a+c)$, $a = a$, $c = c$

    $(-a-c)^2 - 4ac$

    $a^2 + ac + ac + c^2 - 4ac$

    $a^2 + c^2 + 2ac - 4ac$

    $a^2 + c^2 - 2ac$

    $\sqrt{a^2 - 2ac + c}$

    $\sqrt{a^2 - ac - ac + c^2}$

    $\sqrt{a(a-c) - c(a-c)}$

    $\sqrt{(a-c)(a-c)} = a - c$

    ∴ the Equation has real roots

    $a \geq c$

26. Solve $2x^2 + 21x + 52 = 0$

    Discriminant: $21^2 - 4(104)$
    $441 - 416 = 25$

    $2x^2 + 8x + 13x + 52 = 0$
    $2x(x+4) + 13(x+4) = 0$

    $(2x + 13)(x + 4) = 0$

    $2x + 13 = 0$ or $x + 4 = 0$

    $x = -\dfrac{13}{2}$ or $x = -4$

27. Find the possible values of a for which the quadratic equation in $x^2 + (a+2)x + a^2 = 0$ has equal roots

    The discriminant $b^2 - 4ac = 0$

Where $a = 1$, $b = a + 2$ & $c = a^2$

$(a + 2)^2 - 4(1 \times a^2) = 0$

You equate to zero since the equation is said to have two equal roots.

$a^2 + 4a + 4 - 4a^2 = 0$

$-3a^2 + 4a + 4 = 0$

$3a^2 - 4a - 4 = 0$

$16 + 48 = 64$

$3a^2 - 6a + 2a - 4 = 0$

$3a(a - 2) + 2(a - 2) = 0$

$(3a + 2)(a - 2) = 0$

$a = -\frac{2}{3}$ or $a = 2$

$\therefore$ The possible value of a for the equation. $x^2 + (a + 2)x + a^2 = 0$ to have equal roots are $-\frac{2}{3}$ & 2

28. If the roots of the equation $2x^2 + 7x + 3 = 0$ are α and β. Find an equation with integer coefficients, the roots of which are $(2\alpha + \beta)$ and $(2\beta + \alpha)$

$\alpha + \beta = -\frac{7}{2}; \quad \alpha\beta = \frac{3}{2}$

$\alpha = -\frac{7}{2} - \frac{\beta}{1}$

$\alpha = \frac{-7 - 2\beta}{2}$

Substitute α in (2)

$\alpha\beta = \frac{3}{2}$

$\left(\frac{-7 - 2\beta}{2}\right)\beta = \frac{3}{2}$

$$-\beta\left(\frac{7+2\beta}{2}\right) = \frac{3}{2}$$

$$-1\left(\frac{7\beta + 2\beta^2}{2}\right) = \frac{3}{2}$$

$$\frac{7\beta + 2\beta^2}{2} = \frac{3}{2}$$

$$14\beta + 4\beta^2 = -6$$

$$4\beta^2 + 14\beta + 6 = 0$$
$$2\beta^2 + 7\beta + 3 = 0$$

$$49 - 24 = 25$$

$$2\beta^2 + 6\beta + \beta + 3 = 0$$

$$49 - 24 = 25$$

$$2\beta^2 + 6\beta + \beta + 3 = 0$$

$$2\beta(\beta + 3) + 1(\beta + 3) = 0$$

$$(2\beta + 1)(\beta + 3) = 0$$

$$\beta = -\frac{1}{2} \text{ or } \beta = -3$$

If $\beta = -\frac{1}{2}$

$$\alpha = \frac{-7 - 2\beta}{2}$$

$$\alpha = \frac{-7 - 2\left(-\frac{1}{2}\right)}{2}$$

$$\alpha = \frac{-7 + 1}{2}$$

$$\alpha = -\frac{6}{2} = -3$$

If $\beta = -3$

$$\alpha = \frac{-7 - 2\beta}{2}; \quad \alpha = \frac{-7 - 2(-3)}{2}$$

$$\alpha = \frac{-7 - 2\beta}{2} = -\frac{1}{2}$$

$\therefore \alpha = -3$ when $\beta = -\frac{1}{2}$ and vice versa

$$\therefore 2\alpha + \beta = 2(-3) + \left(-\frac{1}{2}\right)$$

$$-\frac{6}{1} - \frac{1}{2}$$

$$= \frac{-12 - 1}{2}$$

$$= -\frac{13}{2}$$

& $2\beta + \alpha = 2\left(-\frac{1}{2}\right) + (-3)$

$$= -1 - 3 = -4$$

The costs are:

$(2\alpha + \beta)$ & $(2\beta + \alpha)$

$(2\alpha + \beta) + (2\beta + \alpha)$

$(2\alpha + \beta) + (2\beta + \alpha)$

$= -\dfrac{13}{2} + \left(-\dfrac{4}{1}\right)$

$= \dfrac{-13 - 8}{2}$

$= -\dfrac{21}{2}$

$(2\alpha + \beta)(2\beta + \alpha)$

$= -\dfrac{13}{2} \times -\dfrac{4}{1}$

$= \dfrac{52}{2} = 26$

The equation is $ax^2 - bx + c = 0$

Where $\dfrac{b}{a} = -\dfrac{21}{2}$

$\dfrac{c}{a} = \dfrac{52}{2}$

$a = 2\ \ b = -21;\ c = 52$

The equation is $2x^2 + 21x + 52 = 0$

29. Find the possible values of a for which the quadratic equation $ax^2 + (a+1)x + a = 0$ has equal roots

The discriminant $b^2 - 4ac$ will be equated to zero i.e. $b^2 - 4ac = 0$

$b^2 - 4ac = 0$

Where $a = a;\ b = (a+1)\ \&\ c = a$

$(a+1)^2 - 4(a \times a) = 0$

$a^2 + 2a + 1 - 4a^2 = 0$

$-3a^2 + 2a + 1 = 0$

$3a^2 - 2a - 1 = 0$

$4 + 12 = 16$

$3a^2 - 3a + a -= 0$

$3a(a-1) + 1(a-1) = 0$

$(3a+1)(a-1) = 0$

$a = -\dfrac{1}{3};\ a = 1$

30. If α and β are the roots of the equation $2x^2 - 5x + 4 = 0$. Find the quadratic equation the roots of which are α^2 and β^2

$\alpha + \beta = \dfrac{5}{2};\ \alpha\beta = \dfrac{4}{2}$

The equation having $\alpha^2\ \&\ \beta^2$ are roots:

$\alpha^2\beta^2 = (\alpha+\beta)^2 - 2\alpha\beta$

$\alpha^2\beta^2 = (\alpha\beta)^2$

$$\alpha^2 + \beta^2 = \left(\frac{5}{2}\right)^2 - 2\left(\frac{4}{2}\right)$$

$$= \frac{24}{4} - \frac{8}{2}$$

$$= \frac{25-16}{4} = \frac{9}{4}$$

$$\alpha^2 + \beta^2 = \frac{9}{4}$$

$$\alpha^2\beta^2 = \left(\frac{4}{2}\right)^2 = \frac{16}{4} = 4$$

If $\alpha^2 + \beta^2 = \frac{9}{4}$ and $\alpha^2\beta^2 = \frac{16}{4}$

The equation $\because$ will be

$ax^2 - bx + c = 0$

Where $\alpha^2 + \beta^2 = \frac{b}{a} = \frac{9}{4}$ & $\alpha^2\beta^2 = \frac{c}{a} = \frac{16}{4}$

$a = 4;\ b = a$ & $c = 16$

The equation is

$4x^2 - 9x + 16 = 0$

31. If one factor of $4t^2 + mt + 1$ is $t - 1$, find m and the other factor

$4t^2 + mt + 1 = 0$

If one of the factor is $(t - 1) = 0$, then $t = 1$

Substitute t in the equation

$4(1)^2 + m(1) + 1 = 0$

$4 + m + 1 = 0$

$5 + m = 0$

$m = -5$

Then $4t^2 + (-5)t + 1 = 0$

$4t^2 - 5t + 1 = 0$

$\dfrac{4t^2 - 5t + 1}{t - 1}$

$$\begin{array}{r}
4t - 1 \\
t - 1 \overline{\smash{)}\, 4t^2 - 5t + 1} \\
-(4t^2 - 4t) \\
\hline
-t + 1 \\
-(-t + 1) \\
\hline
 \cdot \quad \cdot
\end{array}$$

∴ The other factor is $4t - 1$

32. If α and β are the roots of the equation $2x^2 + 6x + k = 0$, find the value of k if $\dfrac{1}{\alpha} + \dfrac{1}{\beta} = -3$

∴ $\alpha + \beta = -\dfrac{6}{2}$

$\alpha\beta = \dfrac{k}{2}$

$\alpha + \beta = -3$

∴ $\dfrac{1}{\alpha} + \dfrac{1}{\beta} = -3$

$$\frac{\alpha+\beta}{\alpha\beta} = -3$$

$$\alpha\beta = \frac{\alpha+\beta}{-3}$$

Where $\alpha+\beta = -3$

$$\alpha\beta = -\frac{3}{-3}$$

$$\alpha\beta = 1$$

If $\frac{k}{2} = \alpha\beta$ then $\frac{k}{2} = 1$

$\therefore k = 2$

33. If the roots of the equation $4x^2 - 6x + 1 = 0$ are α and β, find the equation the roots of which are $\left(\alpha + \frac{1}{\beta}\right)$ and $\left(\beta + \frac{1}{\alpha}\right)$

$4x^2 - 6x + 1 = 0$;

$\alpha + \beta = \frac{6}{4}$ & $\alpha\beta = \frac{1}{4}$

Roots are

$\left(\alpha + \frac{1}{\beta}\right)$ & $\left(\beta + \frac{1}{\alpha}\right)$

$\left(\alpha + \frac{1}{\beta}\right) + \left(\beta + \frac{1}{\alpha}\right)$

$\frac{\alpha}{1} + \frac{1}{\beta} + \frac{\beta}{1} + \frac{1}{\alpha}$

$= \frac{\alpha^2\beta + \alpha + \alpha\beta^2 + \beta}{\alpha\beta}$

$= \frac{\alpha + \beta + \alpha\beta(\alpha+\beta)}{\alpha\beta}$

Substitution

$$\frac{\frac{6}{4} + \frac{1}{4}\left(\frac{6}{4}\right)}{\frac{1}{4}} = \left(\frac{6}{4} + \frac{6}{16}\right) \div \frac{1}{4}$$

$$= \left(\frac{24+6}{16}\right) \div \frac{1}{4}$$

$$= \frac{20}{16} \times \frac{4}{1} = \frac{15}{2}$$

$$\left(\alpha + \frac{1}{\beta}\right)\left(\beta + \frac{1}{\alpha}\right)$$

$$\alpha\beta + 1 + 1 + \frac{1}{\alpha\beta}$$

$$= 2 + \alpha\beta + \frac{1}{\alpha\beta}$$

$$= 2 + \frac{1}{4} + 1 \div \frac{1}{4}$$

$$= 2 + \alpha\beta + \frac{1}{\alpha\beta}$$

$$= 2 + \frac{1}{4} + 1 \div \frac{1}{4}$$

$$\frac{2}{1} + \frac{1}{4} + \frac{4}{1}$$

$$= \frac{8+1+16}{4} = \frac{25}{4}$$

The equation is $x^2 - \left(\frac{15}{2}\right)x + \frac{25}{4} = 0$

$$x^2 - \left(\frac{30}{4}\right)x + \frac{25}{4} = 0$$

$$= 4x^2 - 30x + 25 = 0$$

34. Find the roots of the quadratic equation $q^2 - 9q + 8 = 0$

$b^2 - 4ac$

$(81) - 4(8) = 0$

$81 - 32 = 49$

$q^2 - 8q - q + 8 = 0$

$q(q - 8) - 1(q - 8) = 0$

$(q - 1)(q - 8) = 0$

$q = 1$ or $q = 8$

35. Solve the quadratic equations
    i. $x^2 + 6x + 5 = 0$
    ii. $4x^2 - 21x + 5 = 0$
    iii. $3y^2 + 11y + 8 = 0$
    iv. $y^2 - 9y - 216 = 0$

i. $b^2 - 4ac$
$36 - 20 = 16 = 4^2$

$x^2 + 5x + x + 5 = 0$
$x(x + 5) + 1(x + 5) = 0$

$(x + 1)(x + 5) = 0$

$x = -1$ or $x = -5$

ii. $b^2 - 4ac$

$441 - 80 = 361 = 19^2$

$4x^2 - 20x - x + 5 = 0$
$4x(x - 5) - 1(x - 5)$
$(4x - 1)(x - 5) = 0$

$x = \dfrac{1}{4}$ or $x = 5$

iii. $b^2 - 4ac$
$121 - 96 = 25 = 5^2$

$3y^2 + 3y + 8y + 8 = 0$

$3y(y + 1) + 8(y + 1) = 0$

$(3y + 8)(y + 1) = 0$

$y = -\dfrac{8}{3}$ or $y = -1$

iv. $b^2 - 4ac$
$361 + 864 = 1225 = 35^2$

$y^2 - 27y + 8y - 216 = 0$
$y(y - 27) + 8(y - 27)$

$(y + 8)(y - 27) = 0$

$y = -8$ or $y = 27$

36. Solve graphically:
$2x^2 - 3x = 9$

x	-3	-2	-1	0	1	2	3
$2x^2$	18	8	2	0	2	8	18
$-3x$	9	6	3	0	-3	-6	-9
$y = 2x^2 - 3x$	27	14	5	0	-1	2	9

GRAPH

$x = 3$ or $x = -1.5 = -\dfrac{3}{2}$

37.
   a. Plot the graph of $y = x^2 - 4x + 5$ between the value $x = 0$ and $x = 5$. Do the same for $y = 2 + x$. Hence, solve the equation

   $x^2 - 5x + 3 = 0$ (*)

   b. Solve equation (*) by any convenient algebraic method
   c. Compare your results in (a) and (b) above.

   a.

x	0	1	2	3	4	5
x^2	0	1	4	9	16	25
$-4x$	0	-4	-8	-12	-16	-20
$x^2 - 4x + 5$	5	2	1	2	5	10
$y = 2 + x$	2	3	4	5	6	7

GRAPH

Since $0 = x^2 - 5x + 3 \equiv x^2 - 4x + 5 = 2 + x$

$0 = x^2 - 5x + 3 \equiv x^2 - 4x - x + 5 - 2 = 0$

$0 = x^2 - 5x + 3 = x^2 - 5x + 3 = 0$

The roots is the pint of intersection of the two graphs, $x^2 - 4x + 5$ and $2 + x$

$x \approx 2.7$ or $x \approx 6.2$

  b. $x^2 - 5x + 3 = 0$

    Using formula

$$\frac{-b \pm \sqrt{b^2 - 4ac}}{2a}$$

    $b = -5, a = 1$ & $c = 3$

$$= \frac{5 \pm \sqrt{25 - 12}}{2}$$

$$x = \frac{5 + \sqrt{13}}{2} \text{ or } x = \frac{5 - \sqrt{13}}{2}$$

  c.   Compare your results in (a) and (b) above.

    The result are equal

38. Using the same scale and axes plot the graph of

  $3y = x^2 - 16$ and $y = x - 2$

  Taking values of x from -5 to 5. Read off the co-ordinates of the points of intersection of the line and the curve.

  Find the equation of which the x – co-ordinates are the roots.

x	-5	-4	-3	-2	-1	0	1	2	3	4	5
x^2	25	16	9	4	1	0	1	4	9	16	25
-16	-16	-16	-16	-16	-16	-16	-16	-16	-16	-16	-16
$y = \dfrac{x^2 - 16}{3}$	3	0	-23	-4	-5	-5.3	-5	-4	-23	0	3
$y = x - 2$	-7	-6	-5	-4	-3	-2	-1	0	1	2	3

  GRAPH

The co-ordinates of the points of intersection are -2 & 5. The equation of which the co-ordinates are the roots is

$$\frac{x^2 - 16}{3} = x - 2$$

$$x^2 - 16 = 3x - 6$$
$$x^2 - 3x - 10 = 0$$

39. Solve the simultaneous equations

$2x - 3y + 5 = 0$

$y = 2x^2 - 2x - 1;$

Using graphical method (by drawing graphs from $x = -5$ to $x = 5$)

In equ. (i)

$-3y = -2x - 5$

$$y = \frac{-(2x+5)}{-3}; \quad y = \frac{2x+5}{3} \quad \text{(i)}$$

x	-5	-4	-3	-2	-1	0	1	2	3	4	5
$2x + 5$	-5	-3	-1	1	3	5	7	9	11	13	15
$\frac{2x+5}{3}$	-1.7	-1	-0.3	0.3	1	1.7	2.3	3	3.7	4.3	5
$2x^2$	50	$\frac{3}{2}$	18	8	2	0	2	8	18	32	50
$-2x$	10	8	6	4	2	0	-2	-4	-6	-8	-10
$2x^2 - 2x - 1$	59	$\frac{3}{9}$	23	11	3	-1	-1	3	11	23	39

GRAPH

The points of intersection in the graph are the solution which are $y = 3; x = 2$

$y = 1.2; x = 0.7$

40. A housewife purchases 20 eggs and 2 tins of beans for ₦220. Another housewife buys 35 eggs and 4 tins of beans, at the same unit prices, for ₦425. Calculate the unit prices.

$20 \: eggs + 2 \: tins = ₦220$ (i)
$35 \: eggs + 4 \: tins = ₦425$ (ii)

Multiply (i) by 2 to give (iii), then subtract (ii) from (iii):

$40\ eggs + 4\ tins = ₦440$ (iii)
$35\ eggs + 4\ tins = ₦425$ (ii)
$$\frac{5\ eggs}{5} = \frac{₦15}{5}$$

$1\ egg = ₦3$

Substitute egg in equation (i):
$20\ eggs + 2\ tins = ₦220$

$₦3 \times 20 + 2\ tins = ₦220$

$2\ tins = ₦220 - ₦60$

$2\ tins\ of\ beans = ₦160$
$1\ tin\ of\ beans = \dfrac{₦160}{2} = ₦80$

41. A two digit number exceeds twice the sum of its digits by 7. Find the number if its tens digit is smaller then its units digit also by 7.

    Let the digits by x & y. Since x is tens & y is unit, $10x + y$ will be the two digit umber [e.g. $56 = 10 \times 5 + 6$

    $\therefore 10x + y = 2(x + y) + 7$ (i)
    $x + 7 = y$ (ii)
    $10x + y = 2x + 2y + 7;$

    $10x - 2x + y - 2y = 7$
    $8x - y = 7$ (iii)
    $-8x - 8y = -56$ (iv)

    $7y = 63$
    $y = \dfrac{63}{7} = 9$

    Substitute y in equation (ii)
    $x + 7 = y$
    $x + 7 = 9$
    $x = 9 - 7$
    $x = 2$

    The number is 29

42. The equations $y = x^2 - 3x + 2$ and $y = mx + 4$ have common root. Plot the graph of the quadratic equation and hence, find possible values of the tangent of the linear equation.

x	-5	-4	-3	-2	-1	0	1	2	3	4

x^2	25	16	9	4	1	0	1	4	9	16
$-3s$	15	12	9	6	3	0	-3	-6	-9	-12
2	2	2	2	2	2	2	2	2	2	2
y	42	30	20	12	6	2	0	0	2	6

Roots are $x = 1$ & $x = 2$

GRAPH

∴ The tangent are -4 & -2 for when $m = -4$ & -2; $y = 0$ at 1 & 2 respectively.

POLYNOMIALS

A polynomial of degree n is a function $f(x) = a_n x^n + a_{n-1} x^{n-1} + \cdots + a_i x^i + \cdots + a_1 x + a_0$

Where a_i's are real numbers, $a_n \neq 0$, $0 \leq i \leq n$ and x is an unknown number.

The numbers a_i, $0 \leq i \leq n$ are called constants or scalers or coefficients of the polynomials.

Monomials are one term expression such as $3a^2 bx, \sqrt{2} abc^3, xz^2$. A monomial is the product of a numerical factor (or a coefficient) and one or more unknowns.

Rule: A product of two or more monomials is also a monomial. Similar monomials may be combined

- A polynomial of degree zero is a constant function $f(x) = a_0$.
- Polynomial of degree 1 is a linear function, $y = mx + c = f(x)$, $m \neq 0$
- Polynomial of degree 2 is a quadratic function $f(x) = ax^4 + bx^3 + cx^2 + dx + e$, $a \neq 0$
- Rules: $a_j x^j + b_j x^j = (a_i + b_j) x^j$

$a_j x^j - b_j x^j = (a_j - b_j) x^j$

$(a_j x^j)(b_k x^k) = (a_j b_k x^{j+k})$

If $f(x) = x^4$, then show that $f(x) = f(-x)$ [even function]

If $f(x) = x^3$, show that $f(-x) = -f(x)$

[odd function]

Division of a polynomial by a polynomial $f(x)$ is divided by $ax + b$, the remainder $R = f\left(\dfrac{b}{a}\right)$, which is the value of the polynomial $f(x)$ when $x = -\dfrac{b}{a}$

$f(x) = (ax + b)Q(x) + R$ (i) equate $ax + b$ to 0 $[ax + b = 0, x = -\dfrac{b}{a}]$

Substitute $x = -\dfrac{b}{a}$ and $ax + b = 0$ in equation (i) $f\left(-\dfrac{b}{a}\right) = R.$

- Factor theorem: A linear function $ax + b$ is a factor of a polynomial $f(x)$ if, and only if $f\left(-\dfrac{b}{a}\right) = 0$

If, and only if, means that, starting from $f\left(-\dfrac{b}{a}\right) = 0$, we get $(ax + b)$ is a factor of $f(x)$, then $f\left(-\dfrac{b}{a}\right) = 0$

- Rational (fractional) Root Theorem

If $\dfrac{p}{q}$ is a fractional root of a polynomial

$$f(x) = a_n x^n + a_{n-1} x^{n-1} + \cdots + a_1 x + a_0, \ a_n \neq 0$$, where $a_0, a_1, a_2, \ldots$ are all integers, and if $\dfrac{p}{q}$ has been reduced to its lowest terms, then.

a. p is a factor of the constant term a_0, and
b. q is a factor of the leading coefficient a_n.

Roots of Cubic Equations

- $\alpha + \beta + \delta = -\dfrac{b}{a}, \ \alpha\beta + \delta\alpha + \delta\beta = \dfrac{c}{a}, \ \alpha\beta\delta = -\dfrac{d}{a}$

$[ax^3 + bx^2 + cx + d = 0, \ a \neq 0$

Symmetric identities in $\alpha\beta\delta$

i. $\alpha^2 + \beta^2 + \delta^2 \equiv (\alpha + \beta + \delta)^2 - 2(\alpha\beta + \delta\beta + \alpha\delta)$

ii. $\dfrac{1}{\alpha} + \dfrac{1}{\beta} + \dfrac{1}{\delta} = \dfrac{\alpha\beta + \beta\delta + \alpha\delta}{\alpha\beta\delta}$

iii. $\alpha^2\beta^2\delta^2 \equiv (\alpha\beta\delta)^2$

iv. $\alpha^2\beta^2 + \beta^2\delta^2 + \alpha^2\beta^2 \equiv (\alpha\beta + \beta\delta + \alpha\delta)^2 - 2\alpha\beta\delta(\alpha + \beta + \delta)$

v. $(\alpha + \beta)^2 + (\beta + \delta)^2 + (\delta + \alpha)^2 \equiv 2(\alpha + \beta + \delta)^2 - 2(\alpha\beta + \beta\delta + \alpha\delta)$

vi. $(\alpha + \beta)(\beta + \delta)(\delta + \alpha) \equiv (\alpha + \beta + \delta)(\alpha\beta + \beta\delta + \alpha\delta) - \alpha\beta\delta$

vii. $\alpha^3 + \beta^3 + \delta^3 \equiv (\alpha + \beta + \delta)^3 - 3(\alpha + \beta + \delta)(\alpha\beta + \beta\delta + \alpha\delta) + 3\alpha\beta\delta$

viii. $\alpha^3\beta^3 + \beta^3\delta^3 + \delta^3\alpha^3 \equiv (\alpha\beta + \beta\delta + \alpha\beta)^3 - 3\alpha\beta\delta(\alpha + \beta + \delta)(\alpha\beta + \beta\delta + \delta\alpha) + 3(\alpha\beta\delta)^2$

- Graphical Solution of polynomial equations: To solve a polynomial equation in x graphically we need to make a table of values for the polynomial expression, plot its graph and obtain the (real) roots by reading off the intercept of the graph with the x- axis

Exercise 8.1

1. Let $f(x) = 2x + 1$ and $g(x) = x + 3$

    Calculate $f(g(x))$ and $g(f(x))$, and show that $g(f(x)) \neq f(g(x))$

    $f(g(x)) = f(x + 3) = 2(x + 3) + 1$
    $f(x + 3) = 2x + 7$

$g(f(x)) = g(2x+1) = 2x+1+3$
$g(2x+1) = 2x+4$

$\therefore 2x+7 \not\equiv 2x+4$
$f(g(x)) \not\equiv g(f(x))$

2.
   i. If $f(x) = 3x^2 - 1$, find $f(0)$.

   $f(0) = 3(0)^2 - 1 = 0 - 1 = 1$

   ii. Let $g(t) = t^2 + 4t + 4 \equiv (t+2)^2$, find $g(-2)$
   $g(-2) = (-2)^2 + 4(-2) + 4$
   $= 4 - 8 + 4$
   $= 0$

   iii. In the identity $ax^2 + bx + c \equiv x^2 - 2$

   Find a, b, c.
   If $ax^2 + bx + c \equiv x^2 - 2$, then $ax^2 = x^2$, $bx = 0x$ and $c = -2$

   Then $a = 1, b = 0$ and $c = -2$

   iv. Add $f(x) = x^3 + 3x^2 + 3x + 1$ to $g(x) = x^3 - 3x^2 + 3x - 1$

   $f(x) + g(x) = (x^3 + 3x^2 + 3x + 1) + (x^3 - 3x^2 + 3x - 1$

   $= x^3 + x^3 + 3x^2 - 3x^2 + 3x + 3x + 1 - 1$
   $= 2x^3 + 6x$

   v. Subtract $g(x) = x^3 - 3x^2 + 3x - 1$ from $f(x) = x^3 - 3x^2 + 3x + 1$

   $f(x) - g(x) = (x^3 + 3x^2 + 3x + 1) - (x^3 - 3x^2 + 3x - 1)$

   $= x^3 - x^3 + 3x^2 + 3x^2 + 3x - 3x + 1 + 1$

   $= 6x^2 + 2$

3. Multiply
   i. $(x^2 - 2)$ by $(x^2 + 2)$
   $(x^2 - 2)(x^2 + 2)$
   $= x^4 + 2x^2 - 2x^2 - 4$
   $= x^4 - 4$

   ii. $(x^2 + 2x + 1)$ by $(x - 1)$

   $(x^2 + 2x + 1)(x - 1)$
   $= x^3 - x^2 + 2x^2 - 2x + x - 1$
   $= x^3 + x^2 - x - 1$

4. If $f(x-2) = 3x^2 + x + 5$, calculate $f(-1)$

$f(-1): x - 2 = -1$
$x = -1 + 2$
$x = 1$

$3(1)^2 + 1 + 5 = 9$

Exercise 8.2
1. Obtain the quotient and remainder on dividing.

   i. $3x^2 - 2x + 4$ by $x - 2$

$$\begin{array}{r} 3x + 4 \\ x-2 \overline{\smash{\big)} 3x^2 - 2x + 4} \\ \underline{-(3x^2 - 6x)} \\ 4x + 4 \\ \underline{-(4x - 8)} \\ 12 \end{array}$$

$q(x) = 3x + 4,\ r(x) = 12$

   ii. $4x^2 + 3x^2 - 2x + 1$ by $x^2 + 2x - 1$

$$\begin{array}{r} 4x - 5 \\ x^2+2x-1 \overline{\smash{\big)} 4x^3 + 3x^2 - 2x + 1} \\ \underline{-(4x^3 + 8x^2 - 4x)} \\ -5x^2 + 2x + 1 \\ \underline{-(-5x^2 - 10 + 5)} \\ 12x - 4 \end{array}$$

$\therefore q(x) = 4x - 5,\ r(x) = 12x - 4$

   iii. $x^3 - 1$ by $x - 1$

$$\begin{array}{r} x^2 + x + 1 \\ x-1 \overline{\smash{\big)} x^3 - 1} \\ \underline{-(x^3 - x^2)} \\ x^2 - 1 \\ \underline{-(x^2 - x)} \\ x - 1 \\ \underline{-(x - 1)} \\ \end{array}$$

$\therefore q(x) = x^2 + x + 1,\ r(x) = 0$

   iv. $x^4 - 1$ by $x^2 + 1$

$$x^2 + 1 \overline{\smash{\big)}\begin{array}{l} x^2 - 1 \\ \hline x^4 - 1 \\ -(x^4 + x^2) \\ \hline -x^2 - 1 \\ -(-x^2 - 1) \\ \hline \quad \cdot \quad \cdot \end{array}}$$

$$\therefore q(x) = x^2 - 1, r(x) = 0$$

v. $x^3 + 1$ by $x + 1$

$$x - 1 \overline{\smash{\big)}\begin{array}{l} x^2 - x + 1 \\ \hline x^3 + 1 \\ -(x^3 + x^2) \\ \hline -x^2 + 1 \\ -(-x^2 - x) \\ \hline x + 1 \\ -(x + 1) \\ \hline \quad \cdot \quad \cdot \end{array}}$$

$$\therefore q(x) = x^2 - x + 1, r(x) = 0$$

2. Divide $6x^4 + 7x^3 + 12x^2 + 10x + 1$ by $2x^2 + x + 4$

$$x - 1 \overline{\smash{\big)}\begin{array}{l} 3x^2 + 2x - 1 \\ \hline 6x^4 + 7x^3 + 12x^2 + 10x + 1 \\ \\ -(6x^4 + 3x^3 + 12x^2) \\ \hline 4x^3 + 10x + 1 \\ -(4x^3 + 2x^2 + 8x) \\ \hline -2x^2 + 2x + 1 \\ -(-2x^2 - x - 4) \\ \hline 3x + 5 \end{array}}$$

3. Simplify the expression

$$\frac{2x^2}{x^2 - a^2} + \frac{a}{x + a} + \frac{x}{a - x}$$

$$\frac{2x^3}{x^2 - a^2} + \frac{a}{x + a} + \frac{x}{-1(x - a)}$$

$$\frac{2x^3}{x^2 - a^2} + \frac{a}{x + a} - \frac{x}{x - a}$$

$$\frac{(2x^2 + a(x - a) - x(x + a)}{x^2 - a^2}$$

$$= \frac{2x^2 + ax - ax - a^2 - x^2}{x^2 - a^2}$$

$$= \frac{2x^2 - a^2 - x^2}{x^2 - a^2}$$

$$= \frac{x^2 - a^2}{x^2 - a^2} = 1$$

4. Find the quotient and the remainder when the polynomial $f(x) = x^3 + x^2 - x + 2$ is divided by $(x - 1)$

$$\begin{array}{r}
x^2 + 2x + 1 \\
x - 1 \overline{)x^3 + x^2 - x + 2} \\
\underline{-(x^3 - x^2)} \\
2x^2 - x + 2 \\
\underline{-(2x^2 - 2x)} \\
x + 2 \\
\underline{-(x - 1)} \\
3
\end{array}$$

5. Divide $x^4 + 3x^3 - 9x^2 - 25x - 6$ by $x^2 - x - 6$

$$\begin{array}{r}
x^2 + 4x + 1 \\
x^2 - x - 6 \overline{)x^4 + 3x^3 - 9x^2 - 25x - 6} \\
\underline{-(x^4 - x^3 - 6x^2)} \\
4x^3 - 3x^2 - 25x - 6 \\
\underline{-(4x^3 - 4x^2 - 24x)} \\
x^2 - x - 6 \\
\underline{-(x^2 - x - 6)} \\
\cdot \cdot
\end{array}$$

$\therefore q(x) = x^2 + 4x + 1, \; r(x) = 0$

6. Find the remainder when $f(x) = 3x^3 + 2x^2 + x - 1$ is divided by $g(x) = x - 3$

$$\begin{array}{r}
3x^2 + 11x + 34 \\
x - 3 \overline{)3x^3 + 2x^2 + x - 1} \\
\underline{-(3x^3 - 9x^2)} \\
11x^2 + x - 1 \\
\underline{-(11x^2 - 33x)} \\
34x - 1 \\
\underline{-(34x - 102)} \\
101
\end{array}$$

$\therefore q(x) = 3x^2 + 11x + 34 \; \; r(x) = 101$

7. Divide $x^2 + 2$ by (i) $(x - 1)$ (ii) $(x + 1)$ what do you notice about the remainders in (i) and (ii)?

$$x + 1$$

```
          x² + 2
x - 1  | ────────
       | -(x² - x)
       | ────────
       |   x + 2
       | -(x - 1)
       | ────────
       |     3
```

Remainders are equal

Exercise 8.3

1. If $(x + 1)$ and $(x - 4)$ are factors of the polynomial

$$f(x) = x^3 + px^2 + qx + 12$$

Where p and q are constants, find the values of p and q. Hence, factorize $f(x)$ completely.

[$x + 1 = 0$, $x = -1$; $x - 4 = 0$, $x = 4$]

$f(-1) = 0$ & $f(4) = 0$

$f(x) = x^3 + px^2 + qx + 12$

$f(-1) = (-1)^3 + p(-1)^2 + q(-1) + 12 = 0$

$-1 + p - q + 12 = 0$

$p - q = -12 + 1$

$p - q = -11$

&

$f(4) = (4)^3 + p(4)^2 + q(4) + 12 = 0$

$64 + 16p + 4q + 12 = 0$

$16p + 4q = -76$

$4p + q = -19$

$p - q = -11$ (i)

$4p + q = -19$ (ii)

$5p = -30$

$p = -6$

If $p = -6$ then

$p - q = -11$

$-6 + 11 = q$

$q = 5$

$f(x): x^3 - 6x^2 + 5x + 12$

$[(x+1)(x-4)]$

$x^2 - 4x + x - 4 = x^2 - 3x - 4$

$f(x) \div x^2 - 3x - 4$

$$\begin{array}{r} x - 3 \\ x^2 - 3x - 4 \overline{\smash{\big)} x^3 - 6x^2 + 5x + 12} \\ -(x^3 - 3x^2 - 4x) \\ \hline -3x^2 + 9x + 12 \\ -(-3x^2 + 9x + 12) \\ \hline \cdot \quad \cdot \quad \cdot \end{array}$$

Factors of $f(x)$ are

$(x-1)$, $(x-4)$ and $(x-3)$

2. Find the zeros of the polynomial $p(x) = x^3 - 4x^2 + x - 6$

Value worth testing are
$\pm 1, \pm 2, \pm 3, \pm 6$

$p(1) = 1 + 4 + 1 - 6 = 0$
$p(-1) = -1 + 4 + 1 - 6 = -4$
$p(2) = 8 + 16 + 2 - 6 = 20$
$p(-2) = -8 + 16 - 2 - 6 = 0$
$p(3) = 27 + 36 + 3 - 6 = 60$
$p(-3) = -27 + 36 - 3 - 6 = 0$
$p(6) = 216 + 144 + 6 - 6 = 360$
$p(-6) = -216 + 144 - 6 - 6 = -184$

$\therefore$ the zero are $1, -2, -3 \Rightarrow (x-1)(x+2)(x+3)$

Exercise 8.4

1. The equation $2x^3 + 5x^2 - x - 1 = 0$ has roots α, β, δ.

    Find the equation the roots of which are $\alpha^3, \beta^3, \delta^3$.

    $2x^3 + 5x^2 - x - 1 = 0$; $a = 2, b = 5, c = -1, d = -1$

$$\alpha + \beta + \delta = -\frac{b}{a} = -\frac{5}{2}$$

$$\alpha\beta + \beta\delta + \alpha\delta = \frac{c}{a} = -\frac{1}{2}$$

$$\alpha\beta\delta = -\frac{d}{a} = -\frac{-1}{2} = \frac{1}{2}$$

$$\therefore \alpha^3 + \beta^3 + \delta^3 = (\alpha + \beta + \delta)^3 - 3(\alpha + \beta + \delta)(\alpha\beta + \beta\delta + \alpha\delta) + 3\alpha\beta\delta$$

$$= \left(-\frac{5}{2}\right)^3 - 3\left(-\frac{5}{2}\right)\left(-\frac{1}{2}\right) + 3\left(\frac{1}{2}\right)$$

$$= -\frac{125}{8} - \frac{15}{4} + \frac{3}{2}$$

$$= \frac{-125 - 30 + 12}{8} = \frac{-155 + 12}{8}$$

$$= -\frac{143}{8}$$

$$\alpha^3\beta^3 + \beta^3\delta^3 + \alpha^3\delta^3 \equiv (\alpha\beta + \beta\delta + \alpha\delta)^3 - 3\alpha\beta\delta(\alpha + \beta + \delta)(\alpha\beta + \beta\delta + \alpha\delta) + 3(\alpha\beta\delta)^2$$

$$= \left(-\frac{1}{2}\right)^3 - 3\left(\frac{1}{2}\right)\left(-\frac{5}{2}\right)\left(-\frac{1}{2}\right) + 3\left(\frac{1}{2}\right)^2$$

$$= -\frac{1}{8} - \frac{15}{8} + \frac{3}{4}$$

$$= \frac{-1 - 15 + 6}{8} = -\frac{10}{8}$$

$$\alpha^3\beta^3\delta^3 \equiv (\alpha\beta\delta)^3$$

$$= \left(\frac{1}{2}\right)^3 = \frac{1}{8}$$

The Equation is $8x^2 + 143x^2 - 10x - 1 = 0$

Exercise 8.5

Plot the graph of $y = -2x^2 + x - 1$ what did you notice?

Tables of values:

x	-2	-1	0	1	2	3
y	-12	-5	-2	-3	-8	-17

Discriminant of polynomial:
$b^2 - 4ac$, where $a = -2, b = 1$ & $c = -2$

$b^2 - 4ac = (1)^2 - 4(-2x - 2)$

$= 1 - 16 = -15$

$b^2 - 4ac < 0$

∴ The polynomial has no real roots.

2. Does the graph of $y = (x-2)^3 = (x-2)(x-2)(x-2)$
$(x^2 - 2x - 2x + 4)(x - 2)$
$(x^2 - 4x + 4)(x - 2)$

Tables of Values: $x^2 - 4x + 4$

x	-2	-1	0	1	2	3	4	5
y	16	9	4	1	0	1	4	9

$x - 2$

x	-2	-1	0	1	2	3	4	5
y	-4	-3	-2	-1	0	1	2	3

GRAPH

The point of intersection and root of 2 for both equations

∴ $x = 2$ (three times)

$(x-2)(x-2)(x-2)$

3. Draw the graphs of $y = x^3$ and $y = 24x + 84$ for the same values of x and hence find only real root of the cubic equation $x^3 - 24x - 84 = 0$

Table of values: $x^3, 24x + 84, x^3 - 24x - 84 = 0$

x	-3	-2	-1	0	1	2	3	4	5	6	7
$y = x^3$	-27	-8	-1	0	1	8	27	64	125	216	343
$y = 24x$	-72	-48	-24	0	24	48	72	96	120	144	168

									0	4	8
$y = 24x + 84$	12	36	60	84	108	132	156	180	204	228	252
$y = x^3 - 24x - 84$	-39	-44	-61	-84	-107	-124	-129	-116	-79	-12	91

GRAPH

$x \approx 6.14$

4. Draw the graph of $y = -x^3$

   Tables of values:

x	-2	-1	0	1	2
$y = -x^3$	8	1	0	-1	8

GRAPH

5. Draw the graph of $y = x^3$ and $y = 7x - 6$ on the same areas. Hence find the three roots of the equation $x^3 - 7x + 6 = 0$

   Tables of values: x^3

x	-3	-2	-1	0	1	2	3
$y = x^3$	-27	-8	-1	0	1	8	27

   $7x - 6$

x	-3	-2	-1	0	1	2	3
$7x$	-21	-14	-7	0	7	14	21
$y = 7x - 6$	-6	-27	-20	-13	-6	1	8	15

   GRAPH

   The three roots are the point of intersection of the two equations linked to the x-axis which are: $-3, 1$ and 2
   i.e. $(x + 3)(x - 1)(x - 2)$

Exercise 8.6

1. Multiply $f(x) = (x - 1)$ by $g(x) = x + 2$

   $f(x) \times g(x) = (x - 1)(x + 2)$

   $f(x)g(x) = x^2 + 2x - x - 2$

   $= x^2 + x - 2$

2. If $f(x) = (x - 2)(1 + 4x + 5x^2)$, find $f(-1)$

   $f(x) = x + 4x^2 + 5x^3 - 2 - 8x - 10x^2$

   $f(x) = x + 4x^2 + 5x^3 - 2 - 8x - 10x^2$

   $f(x) = 5x^3 - 6x^2 - 7x - 2$

   $f(-1) = 5(-1)^3 - 6(-1)^2 - 7(-1) - 2$

   $= -5 - 6 + 7 - 2 = -6$

3. If $f(x + 1) = 2x^2 + 3x + 3$, Calculate $f(-1)$

$x + 1 = -1; x = -2$

$f(-1) = 2(-2)^2 + 3(-2) + 3$

$f(-1) = 8 + 6 + 3 = 5$

4. Simplify $(x-3)^3$

$(x-3)^3 = (x-3)(x-3)(x-3)$
$= (x^2 - 3x - 3x + 9)(x-3)$
$= (x^2 - 6x + 9)(x-3)$
$= x^3 - 3x^2 - 6x^2 + 18x + 9x - 27$
$(x-3)^3 = x^3 - 9x^2 + 27x - 27$

5. Find be in the following identity:
$(x+5)(2x-8) - (x+4)(x+9) \equiv x^2 + bx + c$
$(2x^2 - 8x + 10x - 40) - (x^2 + 9x + 4x + 36) \equiv x^2 + bx + c$

$(2x^2 + 2x - 40) - x^2 - 13x - 36 \equiv x^2 + bx + c$

$x^2 - 11x - 76 \equiv x^2 + bx + c$

$bx = -11x$

$b = -11$
$c = -76$

6. Simplify

  i. $(x+1)(x^2 - x + 1)$

  $x^3 - x^2 + x + x^2 - x + 1$
  $= x^2 + 1$

  ii. $(x-1)(x^2 + x + 1)$

  $x^3 + x^2 + x - x^2 - x - 1$
  $= (x^3 - 1)$

  iii. $(x+1)^3$

  $(x+1)(x+1)(x+1)$

  $= (x^2 + x + x + 1)(x+1)$

  $= x^3 + x^2 + 2x^2 + 2x + x + 1$

  $= x^3 + 3x^2 + 3x + 1$

  iv. $(x+1)^4$

  $(x+1)(x+1)(x+1)(x+1)$

$(x^2 + 2x + 1)(x^2 + 2x + 1)$

$x^4 + 2x^3 + x^2 + 2x^3 + 4x^2 + 2x + x^2 + 2x + 1$

$$= x^4 + 4x^3 + 6x^2 + 4x + 1$$

7. If $f(x) = x^3$ and $g(x) = x + 3$

   Find $f(g(x))$ and $g(f(x))$

   $$f(g(x)) = (x + 2)^3$$

   $$f(g(x)) = (x + 2)(x + 2)(x + 2)$$

   $$f(g(x)) = (x^2 + 2x + 2x + 4)(x + 2)$$
   $$= x^3 + 6x^2 + 12x + 8$$

   $$g(f(x)) = x^3 + 2$$

8. Divide $3x^2 - 4x - 4$ by $3x + 2$

   $$\begin{array}{r} x - 2 \\ 3x+2 \overline{\smash{\big)}\, 3x^2 - 4x - 4} \\ -(3x^2 + 2x) \\ \hline -6x - 4 \\ -(-6x - 4) \\ \hline \end{array}$$

   $q(x) = x - 2, \; r(x) = 0$

9. If $f(x) = (3x + 2)(x - 2)$

   Find
   i. $f(2)$

   $$f(x) = 3x^2 - 6x + 2x - 4$$

   $$f(2) = 12 - 12 + 4 - 4$$

   $$f(2) = 0$$

   ii. If $f(x) = (3x + 2)(x - 2)$

   Find $f\left(-\dfrac{2}{3}\right)$

   $$f(x) = 3x^2 - 6x + 2x - 4$$
   $$= 3x^2 - 4x - 4$$

   $$f\left(-\dfrac{2}{3}\right) = 3\left(-\dfrac{2}{3}\right)^2 - 4\left(-\dfrac{2}{3}\right) - 4$$
   $$= -\dfrac{2}{1} - \dfrac{8}{3} - \dfrac{4}{1}$$

   $$= \dfrac{-6 - 8 - 12}{3}$$

   $$= -\dfrac{26}{3}$$

$$= -8\frac{2}{3}$$

10. Find the value of x for which $\dfrac{2x}{x^2 - 2x + 1}$ is not valid.

$\dfrac{2x}{0}$ will make the expression not valid.

$x^2 - 2x + 1 = 0$

$[1x^2; -x - x]$

$x(x - 1) - 1(x - 1) = 0$

$(x - 1)(x - 1) = 0$

$x - 1 = 0$ or $x - 1 = 0$

$x = 1$ or $x = 1$

11. If $f(x) = 4x^3 - 8x^2 - 34x - 4$, find $f(-2)$

$f(-2) = 4(-2)^3 - 8(-2)^2 - 34(-2) - 4$
$= 4(-8) - 8(4) - 34(-2) - 4$
$= -32 - 32 + 68 - 4$
$= -64 - 4 + 68 = 0$

12. If $f(x) = 2x^3 + x^3 - 15x - 18$, Calculate $f(3)$.

$f(3) = 2(3)^3 + (3)^2 - 15(3) - 18$
$= 54 + 9 - 45 - 18$
$= 63 - 63 = 0$

13. If $f(x) = x^3 + 2x^2 - 13x - 10$, find
    i. $f(0)$
    ii. $f(-1)$
    iii. $f(-2)$
    iv. $f(5)$

i. $f(0) = 0 - 0 - 0 - 10$
$= -10$

ii. $f(-1) = -1 - 2 + 13 - 10$
$= 0$

iii. $f(-2) = -8 - 4 - 26 - 10$
$= -48$

iv. $f(5) = 125 - 50 - 65 - 10$
$= 0$

14. Divide $x^4 + 6x^3 + 13x^2 + 12x + 4$ by $x^2 + 3x + 2$

$$\begin{array}{r}
x^2 + 3x + 2 \\
x^2 + 3x + 2 \overline{\smash{\big)} x^4 + 6x^3 + 13x^2 + 12x + 4} \\
- (x^4 + 3x^3 + 2x^2) \\
\overline{ 3x^3 + 11x^2 + 12x + 4} \\
- (3x^3 + 9x^2 + 6x) \\
\overline{ 2x^2 + 6x + 4}
\end{array}$$

$$-(2x^2 + 6x + 4)$$
$$\cdot \quad \cdot$$

15. Find the quotient $q(x)$, and the remainder, $r(x)$ when
    i. $2x^3 + 3x^2 - 1$ is divided by $x^2 - x - 2$
    ii. $3x^3 + 4x^2 - 2x - 1$ is divided by $x^2 + x - 2$

i.

$$\begin{array}{r} 2x + 5 \\ x^2 - x - 2 \overline{\smash{\big)}\ 2x^3 + 3x^3 - 1} \\ -(2x^3 - 2x^2 - 4x) \\ \hline 5x^2 + 4x - 1 \\ -(5x^2 - 5x - 10) \\ \hline 9x - 9 \end{array}$$

$\therefore q(x) = 2x + 5, \ r(x) = 9x - 9$

ii.

$$\begin{array}{r} 3x + 1 \\ x^2 \mp x - 2 \overline{\smash{\big)}\ 3x^3 + 4x^2 - 2x - 1} \\ -(3x^3 + 3x^2 - 6x) \\ \hline x^2 + 4x - 1 \\ -(x^2 + x - 2) \\ \hline 3x + 1 \end{array}$$

$\therefore q(x) = 3x + 1, \ r(x) = 3x + 1$

16. If $f(x) = 2x^2 - 9x - 5$, find
    i. $f(5)$
    ii. $f\left(-\dfrac{1}{2}\right)$

i. $f(5) = 2(5)^2 - 9(5) - 5$
$= 50 - 45 - 5 = 0$

ii. $f\left(-\dfrac{1}{2}\right) = 2\left(-\dfrac{1}{2}\right)^2 - 9\left(-\dfrac{1}{2}\right) - 5$
$= \dfrac{1}{2} + \dfrac{9}{2} - \dfrac{5}{1}$
$= \dfrac{1 + 9 - 10}{2} = 0$

17. Find the zeros of $x^4 - 4x^3 + 4x^2 - 9$ and hence factorize the polynomial completely.

[Value worth testing are: $\pm 1, \pm 3, \pm 9$]

Rational solutions:
$f(1) = 1 - 4 + 4 - 9 = -8$
$f(-1) = 1 + 4 + 4 - 9 = 0$
$f(3) = 81 - 108 + 36 - 9 = -117 + 117 = 0$

$f(-3) = 81 + 108 + 36 - 9 = 72 + 144 = 218$
$f(9) = 81 \cdot 81 - 36 \cdot 81 + 4 \cdot 81 - 9$
$= 81(81 - 36 + 4) - 9$
$= 81(49) - 9$
$= 3939 - 9$
$= 3920$

$f(-9) = 81 \cdot 81 + 81 \cdot 36 + 4 \cdot 81 - 9$
$= 81(81 + 36 + 4) - 9$
$= 81(121) - 9$
$= 9801$

$x - 3, x = -1$
$(x - 3)(x + 1) = x^2 + x - 3x - 3$
$= x^2 - 2x - 3$

$x^4 - 4x^3 + 4x^2 - 9 \div x^2 - 2x - 3$

$$
\begin{array}{r}
x^2 - 2x + 3 \\
x^2 - 2x - 3 \overline{\smash{\big)} x^4 - 4x^3 + 4x^2 - 9} \\
\underline{-(x^4 - 2x^3 - 3x^2)} \\
-2x^3 + 7x^2 - 9 \\
\underline{-(-2x^3 + 4x^2 + 6x)} \\
3x^2 - 6x - 9 \\
\underline{-(3x^2 - 6x - 9)} \\
\cdot \quad \cdot \quad \cdot
\end{array}
$$

$q(x) = x^2 - 2x + 3$

∴ The zeros of $x^4 - 4x^3 + 4x^2 - 9$

are: $(x - 3)(x + 1)(x^2 - 2x + 3)$

18. Factorize completely:

$z^4 - 4z^3 - 2z^2 + 12z + 9$

[Value worth testing are: $\pm 1, \pm 3, \pm 9$]

Rational Solution

$f(1) = 1 - 4 - 2 + 12 + 9 = 16$

$f(-1) = 1 + 4 - 2 - 12 + 9 = 0$

$f(3) = 81 - 108 - 18 + 36 + 9 = 0$

$f(-3) = 81 + 108 - 18 - 36 + 9 = 144$

$f(9) = 81 \cdot 81 - 36 \cdot 81 - 2 \cdot 81 + 108 + 9$
$= 81(81 - 36 - 2) + 117$
$= 81(43) + 117$
$= 3482 + 117 = 3600$

$f(-9) = 81 \cdot 81 + 36 \cdot 81 - 2 \cdot 81 - 108 + 9$
$= 81(81 + 36 - 2) - 108 + 9$
$= 81(115) - 99$
$= 9215 - 99 = 9216$

$z = -1, \ z = 3$

$\therefore \ (z+1)(z-3) = z^3 - 3z + z - 3$
$= z^2 - 2z - 3$

$z^4 - 4z^3 - 2z^2 + 12z + 9 \div z^3 - 2z - 3$

$$
\begin{array}{r}
z^2 - 2z - 3 \\
z^2 - 2z - 3 \overline{\smash{\big)}\ z^4 - 4z^3 - 2z^2 + 12z + 9} \\
-(z^4 - 2z^3 - 3z^2) \\
\hline
-2z^3 + z^2 + 12z + 9 \\
-(-2z^3 + 4z^2 + 6z) \\
\hline
-3z^2 + 6z + 9 \\
-(-3z^2 + 6z + 9) \\
\hline
\cdot \quad \cdot \quad \cdot
\end{array}
$$

$\therefore q(x) = z^2 - 2z - 3$

$\therefore$ the root are
$(z+1)(z-3)(z^2 - 2z - 3)$
$= (z+1)(z-3)(z+1)(z-3)$
$= (z+1)^2(z-3)^2$

19. Factorize $2x^3 + 3x - 5$

[Values worth testing are $\pm 1, \ \pm 2, \ \pm 5, \ \pm \dfrac{5}{2}$]

Rational Solution

$f(1) = 2(1)^3 + 3(1) - 5$

$f(1) = 2 + 3 - 5 = 0$
$f(-1) = -2 - 3 - 5 = -10$
$f\left(\dfrac{1}{2}\right) = 2\left(\dfrac{1}{8}\right) + \dfrac{3}{2} - \dfrac{5}{1}$

$f\left(\dfrac{1}{2}\right) = \dfrac{1}{4} + \dfrac{3}{2} - \dfrac{5}{1}$

$f\left(\dfrac{1}{2}\right) = \dfrac{1 + 6 - 20}{4} = -\dfrac{13}{4}$

$f\left(-\dfrac{1}{2}\right) = -\dfrac{1}{4} - \dfrac{3}{2} - \dfrac{5}{1}$

$f\left(-\dfrac{1}{2}\right) = \dfrac{-1 - 6 - 20}{4} = -\dfrac{27}{4}$

$f(5) = 250 + 15 - 5 = 240$

$f(-5) = -250 - 15 - 5 = -270$

$f\left(\dfrac{5}{2}\right) = \dfrac{125}{4} + \dfrac{15}{2} - \dfrac{5}{1}$

$$= \frac{125 + 30 - 20}{4} = \frac{135}{4}$$

$$f\left(-\frac{5}{2}\right) = -\frac{125}{4} - \frac{15}{2} - \frac{5}{1}$$

$$f\left(-\frac{5}{2}\right) = \frac{-125 - 30 - 20}{4} = -\frac{175}{4}$$

∴ $x = 1$; $(x - 1)$

$2x^3 + 3x - 5 \div (x - 1)$

```
                2x² + 2x + 5
        ┌─────────────────────
x - 1   │ 2x³ + 3x - 5
        │ -(2x³ - 2x²)
        ├─────────────────────
        │ 2x³ + 3x - 5
        │ -(2x³ - 2x²)
        ├─────────────────────
        │ 2x² + 3x - 5
        │ -(2x² - 2x)
        ├─────────────────────
        │ 5x - 5
        │ -(5x - 5)
        ├─────────────────────
        │    .    .
```

∴ The factorized version is:

$(x - 1)(2x^2 + 2x + 5)$

20. Factorize $2x^3 + x - 3$

[Value worth taking $\pm 1, \pm \frac{1}{2}, \pm 3, \pm \frac{3}{2}$]

Rational Solution

$f(1) = 2 + 1 - 3 = 0$

…

∴ $x = 1$; $(x - 1)$

```
                2x² + 2x + 3
        ┌─────────────────────
x - 1   │ 2x³ + x - 3
        │ -(2x³ - 2x²)
        ├─────────────────────
        │ 2x² + x - 5
        │ -(2x² - 2x)
        ├─────────────────────
        │ 3x - 3
        │ -(3x - 3)
        ├─────────────────────
        │    .    .
```

∴ $q(x) = 2x^2 + 2x + 3$

The factorized version is:
$(x - 1)(2x^2 + 2x + 3)$

21. Factorize completely $x^4 - c^4$ where c is a constant.

    [Value worth testing $\pm 1, \pm c$]

    $f(1) = 1 - c^4$

    $f(-1) = 1 - c^4$

    $f(c) = c^4 - c^4 = 0$

    $f(-c) = c^4 - c^4 = 0$

    $x = c$ & $x = c$

    i.e. $(x - c)(x + c)$

    $x^2 + cx - cx - c^2 = x^2 - c^2$

    $$\begin{array}{r} x^2 + c^2 \\ x-1 \overline{\smash{\big)}\, x^4 - c^4} \\ \underline{-(x^4 - c^2 x^2)} \\ c^2 x^2 - c^4 \\ \underline{-(c^2 x^2 - c^4)} \\ \cdot \quad \cdot \end{array}$$

Factorized version: $(x - c)(x + c)(x^2 + c^2)$

22. Factorize completely

    $18y^3 - 9y^2 - 17y - 4$

    [Values worth testing $\pm 1, \pm 2, \pm \frac{2}{3}, \pm \frac{1}{2}, \pm \frac{1}{3}, \pm \frac{1}{6}, \pm 4, \pm \frac{4}{3}$]

    Rational Solution:

    $f(1) = 18 - 9 - 17 - 4 = -12$

    $f(-1) = -18 - 9 + 17 - 4 = -14$

    $f(2) = 144 - 36 - 34 - 4 = 70$

    $f(-2) = -144 - 36 + 34 - 4 = -150$

    $f\left(\frac{1}{2}\right) = \frac{9}{4} - \frac{9}{4} - \frac{17}{2} - \frac{4}{1} = \frac{-17 - 8}{2} = -\frac{25}{2}$

    $f\left(-\frac{1}{2}\right) = -\frac{9}{4} - \frac{9}{4} + \frac{17}{2} - \frac{4}{1}$
    $= \frac{-9 - 9 + 34 - 16}{4} = \frac{-34 + 34}{4} = 0$

...

$\therefore y = -\dfrac{1}{2}; \left(y + \dfrac{1}{2}\right)$ or $2y + 1$

$18y^3 - 9y^2 - 17y - 4 \div 2y + 1$

$$
\begin{array}{r}
9y^2 - 9y - 4 \\
2y + 1 \overline{\smash{\big)}\ 18y^3 + 9y^2 - 17y - 4} \\
\underline{-(18y^3 + 9y^2)} \\
-18y^2 - 17y - 4 \\
\underline{-(-18y^2 - 9y^2)} \\
-8y - 4 \\
\underline{-(-8y - 4)} \\
\cdot \quad \cdot
\end{array}
$$

$\therefore q(x) = 9y^2 - 9y - 4$

Factorized version
$(2y + 1)(9y^2 - 9 - 4)$

23. Find all the zeros of $2t^3 + t^2 - 5t - 3$

[Values worth testing $\pm 1, \pm 3, \pm \dfrac{1}{2}, \pm \dfrac{3}{2}$]

Rational Solution

$f(1) = 2 + 1 - 5 - 3 = -5$

$f(1) = -2 + 1 + 5 - 3 = -45$

$f(-3) = -54 + 9 + 15 - 3 = -33$

$f\left(\dfrac{1}{2}\right) = \dfrac{1}{4} + \dfrac{1}{4} - \dfrac{5}{2} - \dfrac{3}{1}$

$= \dfrac{1 + 1 - 10 - 12}{4} = -5$

$f\left(-\dfrac{1}{2}\right) = -\dfrac{1}{4} + \dfrac{1}{4} - \dfrac{5}{2} - \dfrac{3}{1}$

$= \dfrac{-5 - 6}{2} = -\dfrac{11}{2}$

$f\left(\dfrac{3}{2}\right) = \dfrac{27}{4} + \dfrac{9}{4} - \dfrac{15}{2} - \dfrac{3}{1}$

$= \dfrac{27 + 9 - 30 - 12}{4}$

$= \dfrac{36 - 42}{4} = -\dfrac{6}{4} = -\dfrac{3}{2}$

$f\left(-\dfrac{3}{2}\right) = -\dfrac{27}{4} + \dfrac{9}{4} + \dfrac{15}{2} - \dfrac{3}{1}$

$= \dfrac{-27 + 9 + 30 - 12}{4}$

$= \dfrac{-39 + 39}{4} = \dfrac{0}{4} = 0$

$t = -\dfrac{3}{2}$; $\left(t + \dfrac{3}{2}\right)$ or $(2t - 3)$

$2t^3 + t^2 - 5t - 3 \div 2t - 3$

$x - 1$ $\begin{array}{r} t^2 - t - 1 \\ \overline{\smash)2t^3 + t^2 - 5t - 3} \\ -(2t^3 + 3t^2) \\ \hline -2t^2 - 5t - 3 \\ -(-2t^2 - 3t) \\ \hline -2t - 3 \\ -(2t - 3) \\ \hline \cdot \quad \cdot \end{array}$

$\therefore\ q(x) = t^2 - t - 1$

Factorized version:

$t = -\dfrac{3}{2},\ \dfrac{1 + \sqrt{5}}{2}\ \&\ \dfrac{1 - \sqrt{5}}{2}$

$(2t + 3)(t^2 - t - 1)$

24. Find all the zeros of $4z^3 - 9z^2 + 10z - 2$

[Values worth testing: $\pm 1,\ \pm 2,\ \pm \dfrac{1}{2},\ \pm \dfrac{1}{4}$]

$f(1) = 4 - 9 + 10 - 2 = 3$

$f(-1) = -4 - 9 - 10 - 2 = -25$

$f(2) = 32 - 36 + 20 - 2 = 14$

$f(-2) = -32 - 36 - 20 - 2 = -90$

$f\left(\dfrac{1}{2}\right) = \dfrac{1}{2} - \dfrac{9}{4} + \dfrac{10}{2} - \dfrac{2}{1}$

$= \dfrac{2 - 9 + 20 - 8}{4}$

$= \dfrac{5}{4}$

$f\left(-\dfrac{1}{2}\right) = -\dfrac{1}{2} - \dfrac{9}{4} - \dfrac{10}{2} - \dfrac{2}{1}$

$= \dfrac{-2 - 9 - 29 - 8}{4}$

$= -\dfrac{39}{4}$

$f\left(\dfrac{1}{4}\right) = \dfrac{1}{16} - \dfrac{9}{16} + \dfrac{10}{4} - \dfrac{2}{1}$

$= \dfrac{1 - 9 + 40 - 32}{16} = \dfrac{0}{16} = 0$

$\therefore z = \dfrac{1}{4}$

$\left(z - \frac{1}{4}\right)$ or $(4z - 1)$

$4z^3 - 9z^2 + 10z - 2 \div 4z - 1$

$$
\begin{array}{r}
2x^2 + 2x + 5 \\
x - 1 \overline{\smash{)}\begin{array}{l} 4z^3 - 9z^2 + 10z - 2 \\ -(4z^2 - z^2) \\ \hline -8z^2 + 10z - 2 \\ -(-8z + 2z) \\ \hline 8z - 2 \\ -(8z - 2) \\ \hline \cdot \quad \cdot \end{array}}
\end{array}
$$

$\therefore q(x) = z^2 - 2z + 2$

Factorized version: the zeros of the equation:
$(4z - 1)(z^2 - 2z + 2)$

25. Write down the cubic equation with solution α, β, γ such that
$\alpha + \beta + \gamma = -\frac{14}{5}$

$\alpha\beta + \beta\gamma + \gamma\alpha = -\frac{6}{5}$

And
$\alpha\beta\gamma = \frac{13}{5}$

Cubic Equation: $5x^3 + 14x^2 - 6x - 13 = 0$

26. A polynomial in x of degree two takes the value:

$\frac{1}{p}$ if $x = 0$

$\frac{1}{p+1}$ if $x = 1$

$\frac{1}{p+2}$ if $x = 2$

Find the values of the polynomial when $x = p + 2$

Let the polynomial be $ax^2 + bx + c$

If $x = 0$, then $c = \frac{1}{p}$ (i)

If $x = 1$

$a + b + c = \frac{1}{p+1}$ (ii)

If $x = 2$ then $4a + 2b + c = \frac{1}{p+2}$ (iii)

Substitute $c = \dfrac{1}{p}$ in equation (ii)

$$a + b + \dfrac{1}{p} = \dfrac{1}{p+1}$$

$$a + b = \dfrac{1}{p+1} - \dfrac{1}{p}$$

$$a + b = \dfrac{p - (p+1)}{p(p+1)}$$

$$a + b = \dfrac{p - p - 1}{p(p+1)} \quad \text{(iv)}$$

Substitute c in equation (iii):

$$4a + 2b + \dfrac{1}{p} = \dfrac{1}{p+2}$$

$$4a + 2b = \dfrac{1}{p+2} - \dfrac{1}{p}$$

$$4a + 2b = \dfrac{p - (p+2)}{p(p+2)}$$

$$4a + 2b = \dfrac{p - p - 2}{p(p+2)}$$

$$\dfrac{2(2a+b)}{2} = -\dfrac{2}{p(p+2)} \times \dfrac{1}{2}$$

$$2a + b = -\dfrac{1}{p(p+2)} \quad \text{(v)}$$

Equation (iv) – (v):

$$a + b = -\dfrac{1}{p(p+1)} \quad \text{(iv)}$$

$$-(2a + b = -\dfrac{1}{p(p+2)} \quad \text{(v)}$$

$$-a = -\dfrac{1}{p(p+1)} + \dfrac{1}{p(p+2)}$$

$$a = \dfrac{(p+2) - (p+1)}{p(p+1)(p+2)}$$

$$a = \dfrac{p + 2 - p - 1}{p(p+1)(p+2)}$$

$$a = \dfrac{1}{p(p+1)(p+2)}$$

Substitute a in equation (iv):

$$a + b = -\dfrac{1}{p(p+1)}$$

$$\frac{1}{p(p+1)(p+2)} + b = -\frac{1}{p(p+1)}$$

$$b = -\frac{1}{p(p+1)} - \frac{1}{p(p+1)(p+2)}$$

$$b = \frac{-(p+2)-1}{p(p+1)(p+2)}$$

$$b = \frac{-2-1-p}{p(p+1)(p+2)}$$

Polynomial:
$$\frac{x^2}{p(p+1)(p+2)} - \frac{(3+p)x}{p(p+1)(p+2)} + \frac{1}{p}$$

When $x = p + 2$

$$= \frac{(p+2)^2}{p(p+1)(p+2)} - \frac{(3+p)(p+2)}{p(p+1)(p+2)} + \frac{1}{p}$$

$$= \frac{(p+2)(p+2)}{p(p+1)(p+2)} - \frac{(3+p)(p+2)}{p(p+1)(p+2)} + \frac{1}{p}$$

$$= \frac{p+2}{p(p+1)} - \frac{3+p}{p(p+1)} + \frac{1}{p}$$

$$= \frac{p+2-3-p-p+1}{p(p+1)}$$

$$= \frac{p}{p(p+1)} = \frac{1}{p+1}$$

27. Factorize completely:

$$x^4 - 6x^2 - 7x - 6$$

[Values worth testing: $\pm 1, \pm 2, \pm 3, \pm 6$]

Rational Solution:

$f(1) = 1 - 6 - 7 - 6 = -18$

$f(-1) = 1 - 6 + 7 - 6 = -4$

$f(2) = 16 - 24 - 14 - 6 = -28$

$f(-2) = 16 - 24 + 14 - 6 = 0$

$f(3) = 81 - 54 - 21 - 6 = 0$

$x = -2$ & $x = 3$

$(x+2)(x-3)$

$x^2 - 3x + 2x - 6$

$x^2 - x - 6$

$x^4 - 6x^2 - 7x - 6 \div x^2 - x - 6$

$$
\begin{array}{r}
x^2 + x + 1 \\
x^2 - x - 6 \overline{\smash{\big)}\, x^4 - 6x^2 - 7x - 6} \\
-(x^4 - x^3 - 6x^2) \\
\hline
x^3 - 7x^2 - 6 \\
-(x^3 - x^2 - 6x) \\
\hline
x^2 - x - 6 \\
-(x^2 - x - 6) \\
\hline
\end{array}
$$

$\therefore\ q(x) = x^2 + x + 1$

Factorized version: $(x+2)(x-3)(x^2+x+1)$

28. Find all the roots of $16z^4 - 56z^2 + 32z + 9$

[Values worth testing are $\pm 1, \pm 3, \pm\frac{1}{2}, \pm\frac{1}{4}, \pm\frac{1}{8}, \pm\frac{3}{2}, \pm\frac{3}{4}, \pm\frac{3}{8}, \pm\frac{3}{16}, \pm\frac{9}{2}, \pm\frac{9}{4}, \pm\frac{9}{8}, \pm\frac{9}{16}$]

Rational Solution:

$f(1) = 16 - 56 + 32 + 9 = 1$

$f(-1) = 16 - 56 - 32 + 9 = 63$

$f\left(\frac{1}{2}\right) = 1 - 14 + 16 + 9 = 12$

$f\left(-\frac{1}{2}\right) = 1 - 14 - 16 + 9 = 20$

$$f\left(\frac{1}{4}\right) = \frac{1}{16} - \frac{7}{2} + \frac{8}{1} + \frac{9}{1}$$

$$= \frac{1 - 56 + 128 + 144}{16} = \frac{217}{16}$$

$$f\left(-\frac{1}{4}\right) = \frac{1 - 56 - 128 + 144}{16}$$

$$= \frac{145 - 184}{16} = \frac{39}{16}$$

None of the values tested positively.

$$16z^4 - 56z^2 + 32z + 9 \equiv (pz^2 + qz + r)(hz^2 + gz + k)$$

$$16z^4 \cdots + 9 \equiv hpz^4 + (gp + hq)z^3 + (kp + gq + hr)z^2 + (kq + gr)z + kr$$

$16z^4 = hpz^4$; $hp = 16$ (i)

$gp + hq = 0$ (ii)

$kp + gq + hr = -56$ (iii)

$kq + gr = 32$ (iv)

$kr = 9$ (v)

Find factors of 16 & 9 (hp & kr) and that will fit into the equation $kp + gq + hr = -56$

Factors of 16: $(h \times p)$

$\times 16$ (a)

2×8 (b)

4×4 (c)

Factors of 9: $(k \times r)$

1×9 (d)

3×3 (e)

Testing (a) & (d):

Let $h = 1, p = 16, k = 1, r = 9$

In equation (ii):

$gp + hq = 0$

$16g + 1q = 0$

$q = -16g$

Substitute in equation (3):

$16 + (-16g^2) + 9 = -56$

$25 - 16g^2 = -56$

$25 - 16g^2 = -56$

$25 + 56 = 16g^2$

$\dfrac{81}{16} = \dfrac{16g^2}{16}$

$g = \sqrt{\dfrac{81}{16}}$

$g = \dfrac{9}{4}$

If $g = \dfrac{9}{4}$ then $q = 16 \times \dfrac{9}{4} = 36$

Substitute in equation 4 to check:

$kq + gr$

$1 \times -36 + \dfrac{9}{4} \times 9$

$= -\dfrac{36}{1} + \dfrac{81}{4}$

$= \dfrac{-144 + 81}{4} = -\dfrac{63}{4} \neq 32$

Testing b & d:

Let $h = 2, p = 8, k = 1, r = 9$

In equation (2)

$8q + 2q = 0$

$2q = -8g$

$q = -4g$

In equation (iii):

$8 + (-8g^2) + 18 = -56$

$26 + 56 = 8g^2$

$\sqrt{\dfrac{82}{8}} = g$

$q = -4\sqrt{\dfrac{82}{8}}$

In equation iv:

$-4\sqrt{\dfrac{82}{8}} + \sqrt{\dfrac{82}{8}} \times 9 \neq 32$

Testing c & d:

Let $h = 4, p = 4, k = 1 \,\&\, r = 9$

In equation (ii):

$4g + 4q = 0$

$q = -g$

In equation (iii)

$4g + 4q = 0$

$q = -g$

In equation (iii)

$4 + (-g^2) + 36 = -56$

$40 + 56 = g^2$

$g = \sqrt{96}$

…

Ans in equation (iv) $\neq 32$

But checking the $+ve$ and $(-ve)$ values:

Let $h = 4, p = 4, k = -1 \,\&\, r = -9$

In equation (ii):

$4g + 4q = 0$

$q = -g$

In equation (iii):

$-4 + (-g^2) + (-36) = -56$

$-40 + 56 = g^2$

$\sqrt{16} = g$

$g = 4$

If $g = 4$ then $q = -4$

Check in equation (iv)

$4 + (-36)$

$-36 + 4 = -32$

Then equation is:

$(4z^2 - 4z - 9)(4z^2 + 4z - 1)$

$z = \frac{1}{2}(1 + \sqrt{2})$, $z = \frac{1}{2}(1 - \sqrt{2})$ & $z = -\frac{1}{2}(\sqrt{10} + 1)$, $z = \frac{1}{2}(\sqrt{10} - 1)$

29. Find two values of c which will make $x^2 + cx + 81$ a perfect square.

$b^2 - 4ac = 0$
$c^2 - 4 \times 1 \times 81 = 0$
$c^2 - 324 = 0$
$c^2 = 324$

$c = \pm\sqrt{324}$

$c = +18$ or $c = -18$

30. The roots of a cubic equation α, β, δ are such that $\alpha + \beta + \delta = \frac{3}{2}$

$\alpha\beta + \beta\delta + \delta\alpha = -\frac{1}{2}$

$\alpha\beta + \beta\delta + \delta\alpha = -\frac{1}{2}$

$\alpha\beta\delta = -\frac{7}{2}$

Obtain the equation, the roots of which are

i. $\frac{1}{\alpha^2}, \frac{1}{\beta^2}, \frac{1}{\delta^2}$

ii. $\frac{1}{\alpha^3}, \frac{1}{\beta^3}, \frac{1}{\delta^3}$

i. $\frac{1}{\alpha^2} + \frac{1}{\beta^2} + \frac{1}{\delta^2} = \frac{\alpha^2\beta^2 + \beta^2\delta^2 + \alpha^2\delta^2}{(\alpha\beta\delta)^2}$

$= \frac{(\alpha\beta + \beta\delta + \alpha\delta)^2 - 2\alpha\beta\delta(\alpha + \beta + \delta)}{(\alpha\beta\delta)^2}$

$= \left[\frac{1}{4} + 2\left(\frac{21}{4}\right)\right] \div \frac{47}{4}$

$$= \left[\frac{1}{4} + \frac{21}{2}\right] \times \frac{4}{49}$$

$$= \left[\frac{1 + 42}{4}\right] \times \frac{4}{49}$$

$$= \frac{43}{49}$$

$$\frac{1}{\alpha^2\beta^2} + \frac{1}{\beta^2\delta^2} + \frac{1}{\alpha^2\delta^2} \equiv \frac{1}{\alpha^2\beta^2} + \frac{1}{\beta^2\delta^2} + \frac{1}{\alpha^2\delta^2}$$

$$\equiv \frac{\alpha^2 + \beta^2 + \delta^2}{\alpha^2\beta^2\delta^2}$$

$$= \frac{(\alpha + \beta + \delta)^2 - 2(\alpha\beta + \beta\delta + \alpha\delta)}{(\alpha\beta\delta)^2}$$

$$= \frac{\left(\frac{3}{2}\right)^2 - 2\left(-\frac{1}{2}\right)}{\left(-\frac{7}{2}\right)^2}$$

$$= \left[\left(\frac{3}{2}\right)^2 - 2\left(-\frac{1}{2}\right)\right] \div \frac{49}{4}$$

$$= \left(\frac{9}{4} + \frac{1}{1}\right) \times \frac{4}{49}$$

$$= \left(\frac{9 + 4}{4}\right) \cdot \frac{4}{49} = \frac{13}{49}$$

$$\frac{1}{\alpha^2} \cdot \frac{1}{\beta^2} \cdot \frac{1}{\delta^2} = \frac{1}{(\alpha\beta\delta)^2} = \frac{1}{\frac{49}{4}} = \frac{4}{49}$$

Equation: $49x^3 - 43x^2 + 13x - 4$

ii. $\quad \frac{1}{\alpha^3}, \frac{1}{\beta^3}, \frac{1}{\delta^3}$

$\quad\quad \frac{1}{\alpha^3} + \frac{1}{\beta^3} + \frac{1}{\delta^3}$

$$= \frac{\alpha^3\beta^3 + \beta^3\delta^3 + \alpha^3\delta^3}{\alpha^3\beta^3\delta^3}$$

$$= \frac{\left(-\frac{1}{2}\right)^3 - 3\left(-\frac{7}{2}\right)\left(\frac{3}{2}\right)\left(-\frac{1}{2}\right) + 3\left(-\frac{7}{2}\right)^2}{\left(-\frac{7}{2}\right)^3}$$

$$= \left[\left(-\frac{1}{8}\right) - \frac{63}{8} + \left(\frac{147}{4}\right)\right] \div \left(-\frac{1}{8}\right) - \frac{63}{8} + \left(\frac{147}{4}\right) \div -\frac{343}{8}$$

$$= -\frac{230}{343}$$

$$\frac{1}{\alpha^3} \cdot \frac{1}{\beta^3} + \frac{1}{\beta^3} \cdot \frac{1}{\delta^3} + \frac{1}{\alpha^3} \cdot \frac{1}{\delta^3} \equiv \frac{1}{\alpha^3\beta^3} + \frac{1}{\beta^3\delta^3} + \frac{1}{\alpha^3\delta^3}$$

$$\equiv \frac{\alpha^3 + \beta^3 + \delta^3}{(\alpha\beta\delta)^3}$$

$$= \frac{(\alpha+\beta+\delta)^3 - 3(\alpha+\beta+\delta)(\alpha\beta+\beta\delta+\alpha\beta) + 3\alpha\beta\delta}{(\alpha\beta\delta)^3}$$

$$= \left[\left(\frac{3}{2}\right)^3 - 3\left(\frac{3}{2}\right)\left(-\frac{1}{2}\right) + 3\left(-\frac{7}{2}\right)\right] \div \left(-\frac{7}{2}\right)^3$$

$$= \left(\frac{27}{8} + \frac{9}{4} - \frac{21}{2}\right) \div -\frac{343}{8}$$

$$= \left(\frac{27 + 18 - 84}{8}\right) \times -\frac{8}{343}$$

$$= \frac{39}{343}$$

$$\frac{1}{\alpha^3} \cdot \frac{1}{\beta^3} \cdot \frac{1}{\delta^3} \equiv \frac{1}{(\alpha\beta\delta)^3}$$

$$= \frac{1}{-\frac{343}{8}} = -\frac{8}{343}$$

Equation: $343x^3 + 230x^2 + 39x + 8$

31. Draw the graph x^3

Tables of Values:

x	-2	-1	0	1	2
$y = x^3$	-8	-1	0	1	8

GRAPH

32. Draw the graph of $y = 8x^3 + 10x^2 + 5x + 1$, and hence find the only real root of the cubic equation.
    $8x^3 + 10x^2 + 5x + 1 = 0$

    Tables f Values

x	-2	-1	0	1	2
$8x^2$	-64	-8	0	8	64
$10x^2$	40	10	0	10	40
$5x + 1$	-9	-4	1	6	11
y	-33	-2	1	24	115

    GRAPH

33. Draw the graph of $y = -5x^2$ and $y = 2x - 1$ on the same axes. Hence, find the two solutions of the quadratic equation $-5x^2 - 2x + 1 = 0$

    Tables of Values:

x	-3	-2	-1	0	1	2	3
$-5x^2$	-45	-20	-5	0	-5	-20	-45
$2x - 1$	-7	-5	-3	1	1	3	5

    GRAPH

34. Draw the graph of $y = 3x^2 + 5x + 2$ and hence find the roots of the quadratic equation $3x^2 + 5x + 2 = 0$

    Tables of Values

x	-3	-2	-1	0	1	2	3
$3x^2$	27	12	3	0	3	12	27
$5x + 2$	-13	-8	-3	2	7	12	17
y	14	4	0	2	10	24	44

    GRAPH

    Roots are -1 & -0.66

35. Draw the graph of $y = x^2$ and $y = -x$ on the same axes. Hence, find the two solutions of the equation $x(x + 1) = 0$

    Tables of Values:

x	-2	-1	0	1	2
$y = x^2$	4	1	0	1	4
$y = -x$	2	1	0	-1	-2

    GRAPH

    Roots are -1 and 0

36. Draw the graph of $y = x^3 - 2x^2 - 13x - 10$ between the values of $x = -3$ through zero to $x = +6$ and hence find all the three roots of the equation $x^3 - 2x^2 - 13x - 10 = 0$.

    Tables of Values:

x	-3	-2	-1	0	1	2	3	4	5	6
x^3	-27	-8	-1	0	1	8	27	64	125	216
$-2x^2$	-18	-8	-2	0	-2	-8	-18	-32	-50	-72
$-13x - 10$	29	16	3	-10	-23	-36	-49	-62	-75	-88
y	-16	0	0	-10	-24	-36	-40	-30	0	55

GRAPH

The three roots are:
-2, -1 and 5

SEQUENCE AND SERIES

Definition: A sequence of numbers is a collection of numbers.
$$a_1, a_2, \cdots, a_n, \cdots$$

Which is written in order. a_1 is called the first term, a_2 the second term, a_3 the third term, $\cdots$, a_n the nth term, ...

Sometimes there is a simple relationship between any number and the preceding a succeeding numbers

Examples:
1. The sequence of counting numbers
   $$1, 2, 3, \cdots, n, n+1, \cdots$$

   Where $n = 1, 2, \cdots$

2. $1, 3, 5, 7, \cdots, (2n-1), (2n+1), \cdots$

   Where $a_n = 2n - 1$, $n = 1, 2, \cdots$ is the sequence of odd counting numbers

3. $1, 4, 13, 40, 121, 364, \cdots$ in which the next term can be obtained by multiplying the known term by 3, and adding 1 to the results. We can write down this relationship as

   $$U_{k+1} = 3U_k + 1, \ k = 1, 2, 3, \cdots, U_1 = 1$$

   This type of relationship is called a recurrence relation.

Note: The recurrent relation is not often useful in writing down the nth term of the sequence at once

However, a formula for the nth term may be derived from the recurrence relation.

From the sequence of number $a_1, a_2, \cdots, a_n, \cdots$ we can form another sequence
$$s_1, s_2, \cdots, s_m, \cdots, s_n, \cdots$$

Of partial sums defined as:
$S_1 = a_1$
$S_2 = a_1 + a_2$
$S_3 = a_1 + a_2 + a_3$
$S_4 = a_1 + a_2 + a_3 + a_4$
$\vdots$
$S_m = a_1 + a_2 + \cdots + a_m$
$\vdots$
$S_n = a_1 + a_2 + \cdots + a_n$

$a_1 + a_2 + \cdots + a_n + \cdots$ is called the limiting sum of the series $a_1 + a_2 + \cdots + a_n$.

This implies, for instance, that the sum of the first n terms is given by S_n.

Suppose, for instance, that we wish to find the sum.

$$T = a_{m+1} + a_{m+2} + \cdots + a_n, \ n > m$$

Then, clearly,

$$T = S_n - S_m$$

Arithmetic Progression (A.P.)
The sequence $a, a+d, a+2d, a+3d, \cdots, a+(n-1)d, \cdots$ where a is the first term and any two consecutive numbers differ by a constant d is called an Arithmetic Progression (A.P.). The number d is called the common difference of the A.P.. The recursive relation which defines an A.P. is

$$a_n = a_{n-1} + d$$

Clearly, the nth term of an A.P. is $a_n = a + (n-1)d$

Arithmetic Mean

Given a sequence of numbers $a_1, a_2, \cdots, a_n$ the numbers

$$\frac{a_1 + a_2 + \cdots + a_n}{n}$$ is called the arithmetic mean.

Corollary

For any three consecutive terms of an A.P., the middle term is the arithmetic mean of the other two terms.

Proof that the sum to n terms of an (A.P.) is given by

$$S_n = \frac{n}{2}[2a + (n-1)d]$$

$$= \frac{n}{2}[a + a_n]$$

$$S_n = a + (a+d) + (a+2d) + \cdots + [a+(n-1)d] \quad (1)$$

If S_n is Rewritten, so that $a_n, (a+(n-1)d)$ is written first, and the first (a) is the last value, we have

$$S_n = [a + (n-1)d] + [a + (n+2)d] + [a + (n-3)d] + \cdots + a \quad (2)$$

If equation (1) is added to (2) we have:

$$S_n + S_n = (a + [a+(n-1)d]) + ((a+d) + [a+(n-2)d]) + ((a+2d) + [a+(n-3)d]) + \cdots + ([a+(n-1)d] + a)$$

$$2S_n = [2a + (n-1)d] + [2a + (n-1)d] + [2a + (n-1)d] + \cdots + [2a + (n-1)d]$$

$$\frac{2S_n}{2} = \frac{n[2a + (n-1)d]}{2}$$

$$S_n = \frac{n}{2}[2a + (n-1)d]$$

$$S_n = \frac{n}{2}[a + a_n]$$

Triangular numbers:

The sums $1, 1+2=3, 1+2+3=6, \cdots, 1+2+3+\cdots+n$ are called triangular numbers.

Proof that the nth term of a triangular number is $\frac{n(n+1)}{2}$

Since the nth term is $1+2+3+\cdots+n$ which is S_n and $S_n = \frac{n}{2}[2a+(n-1)d]$ in the term $1+2+3+\cdots+n$

$a = 1$ and $d = 1$

$$\therefore S_n = \frac{n}{2}[2 \times 1 + (n-1)1]$$

$$S_n = \frac{n}{2}[2 + n - 1]$$

$$S_n = \frac{n}{2}(n + 1)$$

$$\therefore 1 + 2 + 3 + \cdots + n = \frac{n(n+1)}{2}$$

Corollary
The sum of two consecutive triangular numbers is a sequence number.

Add n^{th} and $(n+1)^{th}$ terms

$$\frac{n(n+1)}{2} + \frac{(n+1)(n+2)}{2}$$

$$= \frac{2n^2 + 4n + 2}{2} = (n+1)^2$$

Geometric Progression: In the sequence below
$a, ar, ar^2, ar^3, \cdots, ar^n, \cdots$

Where a is the first term and any two consecutive numbers differ by a factor r, is called a Geometric Progression (G.P.). The constant number r is called the common ratio of the G.P.

Since $\dfrac{a_n}{a_{n-1}} = r$ for all n.

The recurrence relation defining a geometric progression is $a_{n+1} = ra_n$.

Geometric Mean:
Given a sequence of n positive numbers $U_1, U_2, \cdots, U_n$ the number $\sqrt[n]{U_1 U_2 \cdots U_n}$ is called their Geometric mean. If x and y are both negative their geometric mean is $-\sqrt{x \cdot y}$

Note: In a G.P. $a_1, a_2, \cdots, a_n, a_{n+1}, a_{n+2}$; a_{n+1} is the geometric mean of a_n and a_{n+2}

i.e. In $a, ar, ar^2, ar^3, \cdots, ar^{n-1}$
Let $a_n = a$, $a_{n+1} = ar$, $a_{n+2} = ar^2$ then a_{n+1} is the geometric mean of a_n and a_{n+2}

$$\sqrt{a_n \cdot a_{n+2}} = \sqrt{a \times ar^2} = ar = a_{n+1}$$

The sum of the first n terms of a G.P. is given by

$$S_n = \frac{a(1 - r^n)}{1 - r} = \frac{a(r^n - 1)}{r - 1}$$

$$S_n = a + ar + ar^2 + \cdots + ar^{n-1} \qquad (1)$$

Multiply S_n by r

$$rS_n = r(a + ar + ar^2 + \cdots + ar^{n-1})$$
$$rS_n = ar + ar^2 + ar^3 + \cdots + a(n-1) \times r$$
$$rS_n = ar + ar^2 + ar^3 + \cdots + (ar^{n-2} \times r) + ar^{n-1+1}$$
$$rS_n = ar + ar^2 + ar^3 + \cdots + ar^{n-1} + ar^n \qquad (2)$$

Equation (1) – (2):

$$S_n - rS_n = a - ar + ar - ar^2 + ar^2 - ar^3 + ar^3 - \cdots - ar^{n-1} + ar^{n-1} - ar^n$$

$$S_n - rS_n = a - ar^n \quad (3)$$

If this then (2) – (1)

$$rS_n - S_n = ar^n - a^1 \quad (4)$$

In equation (3)

$$\frac{S_n(1-r)}{1-r} = \frac{a - ar^n}{1-r}$$

$$S_n = \frac{\left(a(1-r^n)\right)}{1-r}$$

In equation (4)

$$\frac{S_n(r-1)}{r-1} = \frac{ar^n - a}{r-1}$$

$$S_n = \frac{a(r^n - 1)}{r-1}$$

$$\therefore S_n = \frac{a(1-r^n)}{1-r} = \frac{a(r^n - 1)}{r-1}$$

A prime number is a number, greater than 1, which is divisible only by itself and (of course) 1. Two prime number are called TWIN PRIMES if their difference is 2 e.g. (5,7), (11,13), (17,19), (29, 31), (41,43)

Theorem:
The number of primes is larger than any finite number; it is infinite
Proof: (Euclid's classical proof)
Suppose there is just a finite number of primes
$p_1, p_2, \cdots, p_n$.

Let $m = (p_1 \cdot p_2 \cdot \cdots \cdot p_n) + 1$. then either m is prime or it is composite.

If m is prime, we have a contradication because m would be larger than any of the n primes listed earlier. Therefore there exist more than n primes.

On the contrary, if m is complete it must be divisible (exactly) by a prime, p. But this prime divisor p is not $p_1, p_2, \cdots, p_n$ because these leave a remainder of 1 on dividing m.

Hence there must be some other prime p, apart from the n primes listed earlier.

Again, we have a contradiction of the original assumption that there exist only n primes $p_1, p_2, \cdots, p_n$. Since n was chosen arbitrarily, it follows that the number of primes is infinite

Harmonic Progression: the sequence
$$\frac{1}{a}, \frac{1}{a+d}, \frac{1}{a+2d}, \cdots, \frac{1}{[a+(n-1)d]}, \cdots$$

Is called a Harmonic Progression (H.P.)

Clearly, the nth term is $\dfrac{1}{[a+(n-1)d]}$.